**COMPUTER NETWORKS SERIES**

# TOPOLOGY CONSTRUCTION FOR BOOTSTRAPPING PEER-TO-PEER SYSTEMS OVER AD-HOC NETWORKS

# COMPUTER NETWORKS SERIES

**File-Sharing Applications Engineering**
*Luca Caviglione*
20009. ISBN 978-1-60741-594-7

**Topology Construction for Bootstrapping Peer-to-Peer Systems over Ad-Hoc Networks**
*Wei Ding*
2009. ISBN: 978-1-60692-919-3

COMPUTER NETWORKS SERIES

# TOPOLOGY CONSTRUCTION FOR BOOTSTRAPPING PEER-TO-PEER SYSTEMS OVER AD-HOC NETWORKS

## WEI DING

Nova Science Publishers, Inc.
*New York*

Copyright © 2009 by Nova Science Publishers, Inc.

**All rights reserved.** No part of this book may be reproduced, stored in a retrieval system or transmitted in any form or by any means: electronic, electrostatic, magnetic, tape, mechanical photocopying, recording or otherwise without the written permission of the Publisher.

For permission to use material from this book please contact us:
Telephone 631-231-7269; Fax 631-231-8175
Web Site: http://www.novapublishers.com

## NOTICE TO THE READER

The Publisher has taken reasonable care in the preparation of this book, but makes no expressed or implied warranty of any kind and assumes no responsibility for any errors or omissions. No liability is assumed for incidental or consequential damages in connection with or arising out of information contained in this book. The Publisher shall not be liable for any special, consequential, or exemplary damages resulting, in whole or in part, from the readers' use of, or reliance upon, this material. Any parts of this book based on government reports are so indicated and copyright is claimed for those parts to the extent applicable to compilations of such works.

Independent verification should be sought for any data, advice or recommendations contained in this book. In addition, no responsibility is assumed by the publisher for any injury and/or damage to persons or property arising from any methods, products, instructions, ideas or otherwise contained in this publication.

This publication is designed to provide accurate and authoritative information with regard to the subject matter covered herein. It is sold with the clear understanding that the Publisher is not engaged in rendering legal or any other professional services. If legal or any other expert assistance is required, the services of a competent person should be sought. FROM A DECLARATION OF PARTICIPANTS JOINTLY ADOPTED BY A COMMITTEE OF THE AMERICAN BAR ASSOCIATION AND A COMMITTEE OF PUBLISHERS.

**Library of Congress Cataloging-in-Publication Data**
Ding, Wei, 1964-
  Topology construction for bootstrapping peer-to-peer systems over ad-hoc networks / Wei Ding.
    p. cm.
  Includes index.
  ISBN 978-1-60692-919-3 (softcover)
 1. Ad hoc networks (Computer networks) 2. Peer-to-peer architecture (Computer networks) I. Title.
TK5105.77.D56                                                                                      2009
004.6'52--dc22                                                                                2009006461

*Published by Nova Science Publishers, Inc. ✦ New York*

# CONTENTS

| | | |
|---|---|---|
| **Preface** | | vii |
| **Chapter 1** | Introduction | 1 |
| **Chapter 2** | Peer-to-Peer Systems | 5 |
| **Chapter 3** | Previous Works on Bootstrapping in Wired Networks | 17 |
| **Chapter 4** | Previous Works on Structured P2P Systems over MANETs | 31 |
| **Chapter 5** | RAN — an Optimal and Realistic Approach | 35 |
| **Chapter 6** | Algorithms | 47 |
| **Chapter 7** | Simulation | 63 |
| **Chapter 8** | Conclusion | 69 |
| **References** | | 71 |
| **Index** | | 77 |

# PREFACE

As leaders of the decentralization movement, Mobile Ad-hoc Networks (MANETs) and Peer-to-Peer (P2P) systems are hot topics in networking community. Decentralization model puts ordinary network users on the driver's seat, gives them much more control, and stimulates their enthusiasm in active participation. It is believed that decentralization will replace the client/server model as dominant model in the new century.

MANETs and P2P systems share similar theoretical foundations. Both break away from the client/server model using multi-hop multicast. Both are dynamic, highly decentralized, self-organized, and self-healed. However, their levels of real world application are polar apart. Cachelogic reported that in January 2006 P2P traffic accounted for approximately 71% of all Internet traffic. On the other hand, only few MANETs applications have been commercially realized. Remarkable research initiatives in the synergy of P2P systems and MANETs have been sparked by this interesting phenomenon. While most focus on routing, the bootstrapping problem remains indispensable for the transplantation of successful approaches in P2P systems into MANETs.

The crucial problem in bootstrapping is topology construction in P2P overlay layer. In this boof, a novel solution for this problem, i.e. the Ring Ad-hoc Network (RAN) protocol, is introduced. RAN builds effective rings in node ID space of the overlay layer for ring-based P2P systems like Chord, Pastry, and Virtual Ring Routing. With this ring, lengthy stabilization is absolutely unnecessary. It uses only neighbor-based multi-hop primitives. To best of author's knowledge, this is first successful attempt in the area.

*Chapter 1*

# INTRODUCTION

## 1.1. Peer-to-Peer Systems and Mobile Ad-hoc Networks

As shown dramatically by YouTube, the power of decentralization is irresistible. In most areas, it is only a matter of time for the decentralization model to replace the prevalent client/server model.

Although the decentralization model is powerful, not every technology of decentralization is successful. For instance, peer-to-peer (P2P) systems and mobile ad-hoc networks (MANETs), two leading technologies in decentralization, are divergent in their commercialization. In case of P2P systems, welcomed applications succeeded in the market, gained popularity, and attracted active research. Research in turn brought in better applications. This is a virtuous cycle. In the case of MANETs, theoretical research dominated the area for more than a decade, but very little has been transfer into commercial application, if we do not include sensor networks as MANETs. Virtually no application has been widely used in real world except Bluetooth-based MANETs.

P2P systems and MANETs share fundamental homogeneity in many aspects. For example, both communicate by multi-hop messaging. Both are characterized by the absence of network infrastructure. Both are against the framework of central controller and instead rely upon self organization. These inherent similarities imply promising probability of successful synergy and transplantation.

The term "P2P system" is used throughout this chapter. However, it is not semantically differentiated with "P2P network."

## 1.2. Bootstrapping P2P Systems over MANETs

In the synergy of P2P systems and MANETs, a trend has been seen in transplanting achievement in P2P systems into MANETs. In this direction, majority of research has focused on issues of stable status, especially routing. Transplantation hence concentrated on layer substitution which match layer model in wired IP networks to layer model of MANETs. [HGRW2006, LLS2004, HPD2003, PDH2004]

Very limited research has been done in exploring the bootstrapping. Bootstrapping has been largely circumvented using unrealistic assumptions. This has been a repeated characteristic in the research of P2P systems over wired networks. [RD2001, SMKKB2001, CCNOR2006]

Bootstrapping includes two major tasks. The first task is automatic nodes address configuration. If we follow the traditional layer model of MANETs and keep the stiff separation between layers, we need two configurations: the lower in the networking layer and the higher in overlay or application layer. The second task is setting up overlay topology. This chapter focuses on the second task.

In computer networks, topology is frequently used to define qualitative geographic relationships, such as "which node is directly connected to which node," or "which node is neighbor of which node." Certain type of structured P2P systems imposes particular topologies among nodes to form a specific global structure. For a structured P2P system, overlay topology is crucial to its functionality. It lays foundation for other functions like routing, resource sharing, advertising, looking up, retrieval, and data dissemination. It is one of dominant factors that affect primary performance parameters such as efficiency, robustness, scalability, and feasibility. In fact, topology has broader functionality. For example, Jelasity and Babaoglu [JB2005] have shown that problems such as clustering and sorting can be transformed into topology problems and be solved by specific topology construction.

## 1.3. Current Status

There are two major approaches for bootstrapping a structured P2P system over wired networks. One is to jumpstart a network from one or a few predefined nodes, in which the common way to expand the network is node joining. In wired networks, many structured P2P systems require manual creation of a "seed" network in bootstrapping. Nodes have to be booted one by one in a slow, linear manner, which costs long time. In addition, the jumpstarted network often needs extra long time for stabilization before normal routing could work. In another approach all nodes cooperate concurrently to construct an overlay topology. This approach is distributed and decentralized. The concurrency makes it much faster than joining approach. Furthermore the P2P system could advance to normal working status immediately after bootstrapping.

Remarkable advance has been recently made in topology construction in wired networks. Topology generators can construct topologies such as line, ring, mesh, star, and tree. Generic generator, which could construct any topology if given a mathematical expression, is already available. [JB2005] However, no such construction tool has been reported in MANETs.

Many problems remain unsolved in this area. In wired networks, existing protocols for topology construction and maintenance are usually based upon unrealistic assumptions. Most of them assume the existence of a specific initial topology. Some protocols demand that the network remains in an ideal topology all the time as a necessary condition for normal operations. The second problem is the ignorance of network merger and partition. Some systems even require each node keep and monitor global state of the entire network. Another problem is: many schemes for topology construction still employ centralized strategy. Some require central coordinators; some follow a network-wide top-down view in protocol design. Some have deficiencies in fault-tolerance and recovery. For structured P2P systems over MANETs, some approaches can not keep up with the rate of change.

## 1.4. RAN – A New Solution

In this chapter, the Ring Ad-hoc Network (RAN) protocol is introduced. It builds a ring topology in P2P ID space. To best of author's knowledge, it is the first successful attempt in topology construction for MANETs. The ring is used to bootstrap ring-based P2P systems, such as Chord [SMKKB2001, DBKKMSB2001], Pastry [RD2001], and Virtual Ring Routing [CCNOR2006], over MANETs. On this ring, ring-based P2P systems could be put into normal operation immediately without lengthy stabilization. As in many literatures, Chord is used in demonstration. Three patterns are tested in RAN: distributed exhaustive pattern, virtual centralized exhaustive pattern, and random pattern. Simulation shows that the distributed exhaustive pattern has best overall performance. So this pattern is essentially representative of RAN.

The rest of this chapter is organized as follows. Section 2 introduces P2P systems, especially Chord. Section 3 depicts three previous research projects —

T-Man, T-Chord, and Ring Network. They are most successful approaches for Chord ring construction in wired networks. Section 4 describes P2P systems in MANETs. Section 5 outlines the RAN protocol suite. Section 6 gives algorithms of RAN in AP notation. Simulation results are given in Section 7. Section 8 concludes the chapter.

*Chapter 2*

# PEER-TO-PEER SYSTEMS

## 2.1. Peer-to-Peer Paradigm

P2P overlay systems provide fast, accurate, and scalable resource discovery, resource sharing, and storage services without a central controller. The concept of P2P systems first appeared in mid 1990s. As file sharing platforms, especially to distribute music over Internet, P2P systems became a hot topic in the late 1990s. A traditional P2P system is built upon IP. It uses IP as the communication platform. An IP capable host can reach anything attached to the Internet or other IP networks like IEEE 802 family by an IP address. However, IP layer could not tell a host how and where to find given content or another host. This is done by P2P overlay systems. The basic task of P2P overlay systems is to connect to other peers and find out interesting content.

P2P systems are distributed and self-organized. A host is called peer, because all hosts usually have same status, share same responsibility, and the relationship among them is characterized by equality. Unused bandwidth, storage, CPU cycles are shared among peers. Peers enjoy great freedom and privacy. Usually consumers are also producers, so aggregate resources grow exponentially with utilization. P2P systems have excellent fault tolerance, because there is no single point of failure in a P2P system.

The emergence of P2P systems was a revolution against long time dominance of client/server model in computing and communication. In the client/server model, powerful, reliable servers provide data and services. Clients request data and services from servers. The client/server model has proved extremely successful by its famous offspring, such as World Wide Web, database systems, and FTP. However, it has following inherent defects:

- need central controller
- dictation in which clients look like slaves
- presents a single point of failure
- unused resources through out the network
- poor scalability

## 2.2. Peer-to-Peer Systems

P2P systems address above defects of client/server model. At large P2P computing aims at sharing and exchanging resources and services between terminals. These resources and services include information (file or data structure), CPU cycles, storage (memory, cache, and disk), I/O devices, etc. P2P paradigm takes advantage of superfluous computing capacity, storage, and network bandwidth, so end users can unite and leverage their collective power to carry out huge task or achieve mutual benefits.

In a P2P system, all nodes are clients, servers, and routers at same time. All provide and consume data and services. No centralized data source endangers the system as the single point of failure. Nodes collaborate directly with each other. Any node can initiate a connection. All nodes are totally free: they may enter and leave the network arbitrarily and frequently. It will be "the ultimate form of democracy on the Internet" as well as "the ultimate threat to copy-right protection on the Internet." [Kaashoek2003]

P2P systems have following advantages: [Muthusamy2003]

- Efficient use of resources
- Unused bandwidth, storage, CPU cycles at the edge of the network become available to any user
- Scalability
    - Consumers of resources also donate resources. If remarkable consumers turn into producer, aggregate resources will grow with utilization.
    - Self-scaling
- Reliability
    - No single point of failure
    - Geographic distribution

- Replicas
- Built-in fault tolerance
- Fault tolerance
- Easy administration
  - Nodes self organizing
  - No need to deploy servers
  - Load balancing

Besides file sharing, P2P paradigm could be applied in collaborative Internet (e.g. ICQ, shared whiteboard), distributed computing and grid computing (e.g. UC Berkley Seti@Home Project), multiplayer network games (e.g. Doom) and many other fields. However, P2P systems, especially those for file sharing, remain to be the oldest and most sophisticated P2P application. In a typical file sharing network, a user makes files (music, video, etc.) on her computer available to others. Then another user connects to the network, searches for the files, finds the first user's computer, and downloads files directly from first user's computer.

P2P systems fall into two categories: unstructured P2P systems and structured P2P systems. [Muthusamy2003] An unstructured P2P system does not have a fixed topology for routing. By the existence of central index servers, unstructured P2P systems are divided into three subgroups: centralized with a central index server, like Napster; semi-centralized with local index servers, like KaZaA; decentralized without any index server, like Gnutella. [Clip2, Ivkovic2001]

Structured P2P systems use fixed topologies like ring or grid for routing. They impose specific local relationships between peers, which finally generate global structures. These topology structures can be used for efficient data placement, search, and retrieval. They have guaranteed scalability — hops in routing is not linear with number of nodes. Most of them could reach the logarithm. They are self-organized, fault-tolerant, and they support load balancing. Structured P2P systems are usually implemented via Distributed Hash Table (DHT). Typical systems include Chord [DBKKMSB2001, SMKKB2001], Pastry [RD2001], CAN [RFHKS2001], BitTorrent [Cohen2003I, Cohen2003B], and Virtual Ring [CCNOR2006].

## 2.3. Unstructured P2P Systems

Napster was devoted to sharing music files on Internet. Providers upload their list of files and IP addresses to Napster server. Downloaders send queries to Napster server for files of their interest in the format of keyword search. Keywords could be artist, song, album, even bit rate. Napster server replies with IP address of users with matching files. Downloaders connect directly to the provider's computer to download file. Using a central directory/index server and a central query database, Napster guarantees correct results. At same time, the central server forms a single point of failure and bottleneck for scalability. Napster is Susceptible to denial of service attack and mischief from malicious users.

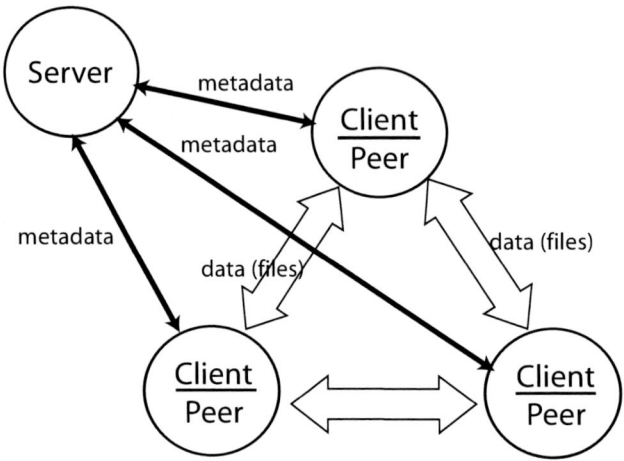

Figure 1. Centralized architecture of Napster

Gnutella enables sharing any type of files, not just MP3. It employs decentralized search. In Gnutella a user A asks her neighbors for files of interest, those neighbors ask their neighbors, and so on. Finally either users with matching files reply to A's query, or the packet is destroyed after a preset Time To Live (TTL). Each message has a parameter which sets the max number of hops the packet can "live". Search is distributed by the means of queries flooding. Comparing to Napster, Gnutella is decentralized and robust to denial of service attacks, for it has no single point of failure. Nevertheless, it can not guarantee correct results for every query. Gnutella is still not scalable. [Clip2, Ivkovic2001]

KaZaA is a hybrid of centralized and decentralized structures, where super-peers act as local central nodes and local search hubs. Each super-peer is similar to a Napster server in a smaller scale. Super-peers are automatically chosen by the system based on their capacities (storage, bandwidth, etc.) and availability (connection time). Users upload their list of files to a super-peer, which periodically exchange file lists with neighbor super-peers. When the query reaches a super-peer for files of interest, the file is transferred back to requesting node following the reverse path. [Muthusamy2003]

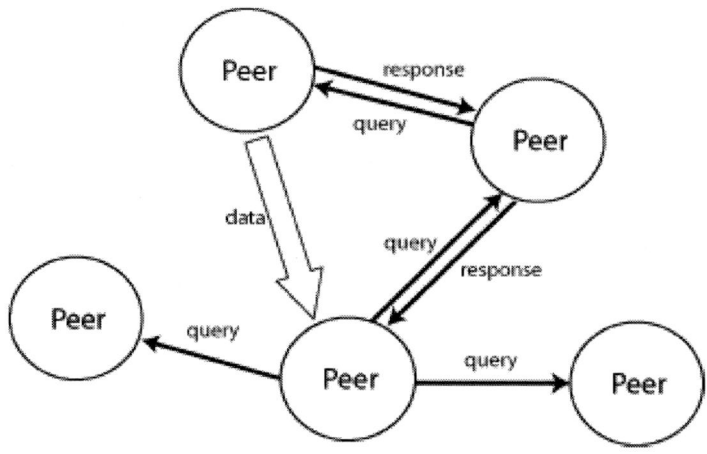

Figure 2. Flooding style search in Gnutella

## 2.4. Structured P2P Systems

Structured P2P systems are also called second generation P2P overlay networks. [Muthusamy2003] They are self-organized and fault-tolerant with balanced load. Scalability is guaranteed on numbers of hops to answer a query. One frequently cited difference with unstructured P2P systems their DHT interface.

A DHT stores (key, value) pairs. Each peer stores a subset of (key, value) pairs. Core functions of DHT API include insert, lookup, and delete. Insert function stores a (key, value) pair at the node responsible for the key. Lookup function returns value associated with a key from the host peer of the pair. Basic operation is to find node responsible for a key. A key need to be mapped to a node before insert, lookup, or delete functions could be used for this node. DHT maps Keys evenly to all nodes in the network. Each node maintains information about only a few other nodes. Messages can be routed to a node efficiently. Arrival or departure of one node only affects a few nodes.

Many services can be built on top of a DHT interface, like file sharing, archival storage, databases, naming, service discovery, chat, rendezvous-based communication, publish and subscribe. There are several implementations of DHT generic interface, for instance, Chord from MIT, Pastry from Microsoft Research in UK and Rice University, Tapestry from UC Berkeley, Content Addressable Network (CAN) also from UC Berkeley, SkipNet from Microsoft Research and University of Washington, Kademlia from New York University,

Viceroy from Israel government and UC Berkeley, P-Grid from EPFL in Switzerland, Freenet developed by Ian Clarke. These systems are also called P2P routing substrates.

Routing in Chord is based upon a ring, on which nodes are organized according to their node IDs. Keys are assigned to their successor node in the ring. The consistent hash function ensures even distribution of nodes and keys on the ring. In a system with $N$ nodes and $K$ keys, lookups are resolved with $O(\log N)$ hops as well.

Pastry has a similar interface to Chord, however, it has good network locality to minimize hop traveling distance. To achieve locality new node needs to know a nearby node. Each routing hop matches the target identifier by one more digit. There are many choices in each hop, called possible locality.

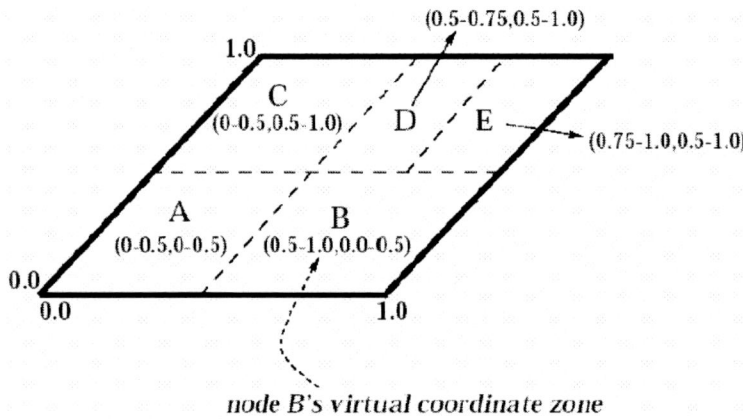

Figure 3. CAN network with 5 nodes in 2-d space

CAN uses a $d$-dimensional Cartesian coordinate space on a $d$-torus. Each node occupies a distinct zone in the space. Each key is hashed to a point in the space.

BitTorrent has a highly connected ring topology with a center, like a bike wheel. BitTorrent uses economic methods in file sharing. It is faster and more reliable than most P2P approaches. BitTorrent forces concurrent downloaders of a same file to share the cost of upload. By using BitTorrent, they have to upload pieces of the file to each other. [Cohen2003I, Cohen2003B]

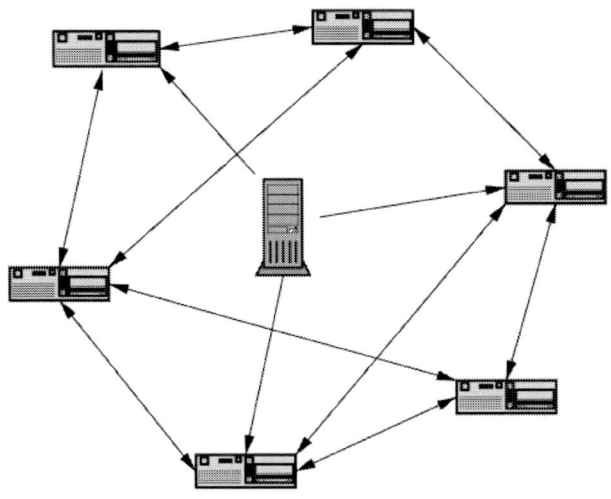

Figure 4. BitTorrent nodes upload pieces of common file to each other

## 2.5. Chord

### 2.5.1. Consistent Hashing

Chord employs consistent hashing to assign ID to nodes and keys. The consistent hashing uses SHA-1 cryptographical hash as its base hash function. The compositive effect of two functions provides fast distributed hash computation. The consistent hashing has three attractive idiosyncrasies.

First, like other DHT, consistent hashing helps routing in Chord remain scalable to network size, that is, node number in the network. Unlike many proactive routing algorithm, Chord does not need its nodes keep tracking of every other node. A Chord node just need track $O(\log N)$ other nodes in its finger table. Each node resolves the hash function by communicating with other nodes. A lookup search for a key in Chord DHT only requires $O(\log N)$ messages to be exchanged.

Second, it has superb load balancing and map keys evenly to nodes with uniform random distribution. This character is very important to Chord's success. It provides solid foundation for Chord's scalability, that is, the scalability to base. Many calculations in Chord involve modular operation. The scalability to base makes Chord calculations independent of base. No matter how big a base you chose, this feature will keep Chord at similar performance level.

Third, consistent hashing is very stable. With help of consistent hashing, Chord could smoothly absorb disturbance from joining and ungraceful leaving (leaving without handling problems arising from the leave). In Chord ID space, a joining or leaving node only affects $O(1/N)$ existing keys in network which need move to other nodes to maintain the network-wide load balance. This is almost theoretical optimum.

### 2.5.2. Routing in Chord

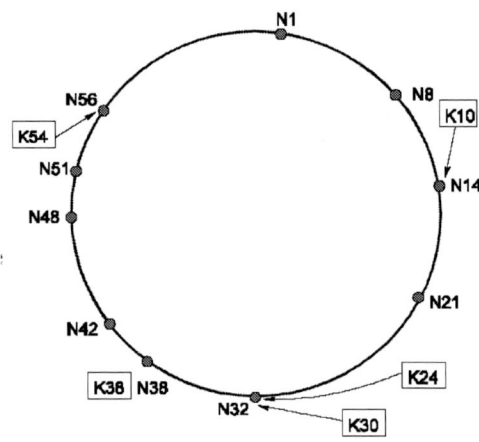

Figure 5. Chord identifier ring.

Routing in Chord is implemented by Chord identifier ring, as shown in Figure 5 and 6, on which nodes are organized according to node IDs. Keys are assigned to their successor node in the ring. The Hash function ensures even distribution of nodes and keys. In an $O (\log N)$ size Chord finger table associated with an $N$ size node set, $i$th finger points to the first node that succeeds $n$ by at least $2^{i-1}$.

To look up a key n, we first locate the furthest node that precedes the key in the finger table. Chord queries could find the target's home address in $O (\log N)$ hops. In a system with $N$ nodes and $K$ keys, with high probability, each node receives at most $K/N$ keys. Each node maintains information about $O (\log N)$ other nodes. And lookups are resolved with $O (\log N)$ hops. However, the efficiency comes with a loss in accuracy. In Chord, there is no guaranteed delivery and no guaranteed consistency among replicas. Hops have poor network locality, nodes close on ring can be amazingly far in the physical network.

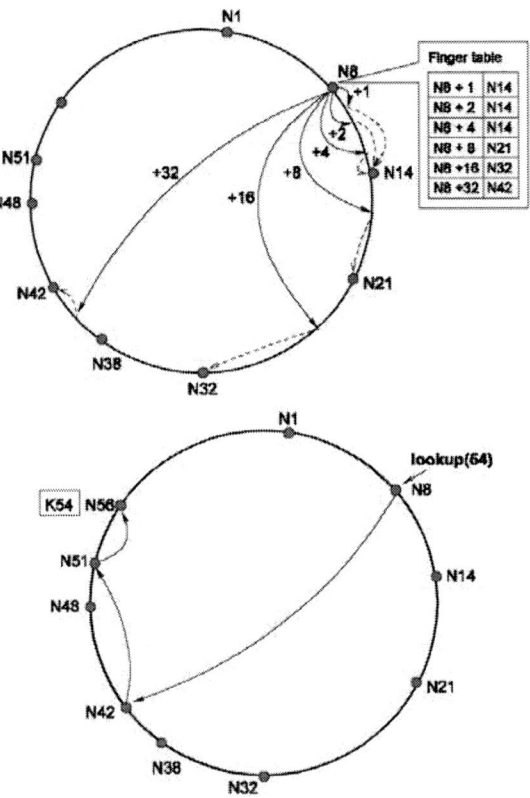

Figure 6. Looking up a key in Chord

## 2.5.3. Chord Algorithm

In Chord, a node ID is a unique m-bit identifier, hashed from IP address or other unique ID. A key is an *m* bit identifier, hashed from a sequence of bytes. A value is sequence of bytes. Chord API includes following functions:

// node *n* finds the successor of *id*
*n*.**find_successor**(*id*)
    **if** (*id* ∈ (*n*, *successor*])
        **return** *successor*;
    **else**
        *p* = closest_preceding_node(*id*);

**return** *p*.find_successor(*id*);

// search the local table for the highest predecessor of *id*
*n*.**closest_preceding_node**(*id*)
 **for** *i* = *m* **downto** 1
  **if** (*finger*[*i*] ∈ (*n*, *id*))
   **return** *finger*[*i*];
 **return** *n*;

// create a new Chord ring
*n*.**create**()
 *predecessor* = nil;
 *successor* = *n*;

// join a Chord ring containing node *p*.
*n*.**join**(*p*)
 *predecessor* = nil;
 *successor* = *p*.find_successor(*n*);

// called periodically. verifies *n*'s immediate successor, and tells the successor about n.
*n*.**stabilize**()
 *x* = *successor.predecessor*;
 **if** (*x* ∈ (*n*, *successor*))
  *successor* = *x*;
 *successor*.notify(*n*);

// n thinks p might be predecessor.
*n*.**notify**(*p*)
 **if** (*predecessor* is nil **or** *p* ∈ (*predecessor*, *n*))
  *predecessor* = *n*;

// called periodically. It refreshes finger table entries. next stores the index of the next finger *to fix*.
*n*.**fix_fingers**()
 *next* = *next* + 1;
 **if** (*next* > *m*)
  *next* = 1;
 *finger*[*next*] = find_successor($n + 2^{next-1}$);

// called periodically. It checks whether predecessor has failed.
*n*.**check_predecessor**()
    **if** (*predecessor* **has failed**)
        *predecessor* = nil;

*Chapter 3*

# PREVIOUS WORKS ON BOOTSTRAPPING IN WIRED NETWORKS

## 3.1. T-Man — A Gossip-Based Approach

Based upon popular gossip communication model [LMM2000] in distributed computing, T-Man [JB2005] is designed as a general purpose protocol for building and maintaining network topology. T-Man targets large scale and highly dynamic networks. It is simple, scalable, robust, and flexible. It may be used as a standalone program, a bootstrapping component, or a recovery component in other protocols. It is mainly used in P2P community, but has an application range far beyond. With the aid of its original concept — the ranking function, T-Man controls self-organization of topologies in a straightforward, intuitive, and adaptive manner. T-Man follows a stepwise refining procedure with a short asymptotic time. T-Man is completely distributed. Each node relies solely upon local communication to increase the quality of the current set of neighbors. Its fast convergence and high robustness in dynamic environments have attracted considerable follow-up research.

T-Man is so adaptive and flexible that it allows for topology change on-the-fly at run time without any change in protocols. All previous approaches have to revise protocol for each possible topology to achieve the same objective. As a general abstraction, topology can be used to solve problems or to enhance and support other solutions. Therefore changing topology on-the-fly will have significant benefit in both theory and practice. It may drastically increase the efficiency of distributed applications as well as the efficiency in deploying such applications. With the support for quick topology change, we can derive best topology for a certain scenario by progressive evolution of topologies.

### 3.1.1. Gossip Protocol

The Gossip protocol [BEGH2004, JHB2001, LMM2000, MMA2000] provides a scheme for performing probabilistically reliable network broadcasts. In the Gossip protocol nodes send a message to some instead of all neighbors (usually only one). The recipients are often selected randomly, but deterministic algorithms are used as well. Due to the redundancy in links, most nodes received the packet in limited hops. Gossip minimizes amount of transportation, and hence reduce communication overhead. Gossip has much better performance than flooding. Gossip can be used to deliver multicast messages with less overhead and enhanced efficiency than normal flooding style broadcasting.

### 3.1.2. Ranking Function

Key concept of T-Man is *ranking function*, which specifies the preference for a node to choose its neighbors in the target topology. A node uses the ranking function to order any set of nodes according to the preference. This simple abstraction results in an effective algorithm which generates various topologies with preciseness and efficiency. The ranking function is the source of effectiveness, versatility, and flexibility of T-Man.

Suppose a network contains nodes, all connected to each other. Each node has an address sufficient for sending messages to it. Each node maintains addresses of other nodes through a partial view, which is a set of node descriptors. Besides a node address, a node descriptor contains a profile, which contains topology related properties, such as ID, geographical location, etc. Links of topology are determined by addresses in partial views descriptors.

Following the selected ranking function, T-Man use local gossip messages and gradually evolves the current topology towards the target. According to its simulation report, the convergence is fast and scalable. Convergence time grows as the logarithm of the network size. The high speed guarantees that T-Man can build divergent topologies on-the-fly. This feature makes T-Man a perfect fit for dynamic systems in which the nodes and their properties change rapidly.

Here gives a formal description. Suppose $N$ is the node set of a network. Each node x maintains addresses of other nodes through a partial view, denoted as

$view_x$. $c$ is the maximal size of partial views in the network. Ranking function $R$ has following parameters as its input.

- $x$, base node
- $S = \{y_1, y_2, ..., y_m\}$, a set of nodes

The output of $R$ is an m-tuple, which is a re-ordered $S$. The task is to construct views of all nodes such that the view of node $x$, $view_x$, contains exactly the first $c$ elements of $R(x, \{$all nodes except $x\})$, which is output of $R$ over the entire node set. That is,

$$R(x, view_x) = R(x, N - \{x\})$$

One convenient way to get a ranking function is through a distance function, which is derived from a metric space over the node set. The ranking function measures the Euclidean or other distance from the base node. Here are few examples of defining distance function. For lines, the profile of a node is a real number. The distance function is

$$d(a, b) = |a - b|$$

Its variant can be extended to a ring. For example for a Chord ring with range $[0, N]$, node profile is an integer in $[0, N]$. Here distance is directional, that is, $d(a, b)$ is not necessarily equal to $d(b, a)$. The distance function is defined as

$$d(a, b) = (a - b) \bmod (N+1)$$

Extending one dimensional distance function for line to two dimensions, we can derive distance function for a mesh. The profile for node is two-dimensional real vector. The distance for the mesh is the Manhattan distance, which is the sum of two one dimensional distances on two coordinates. Use the same transformation from line to ring, we can get profile and distance function for tube from those for mesh.

### 3.1.3. T-Man Protocol

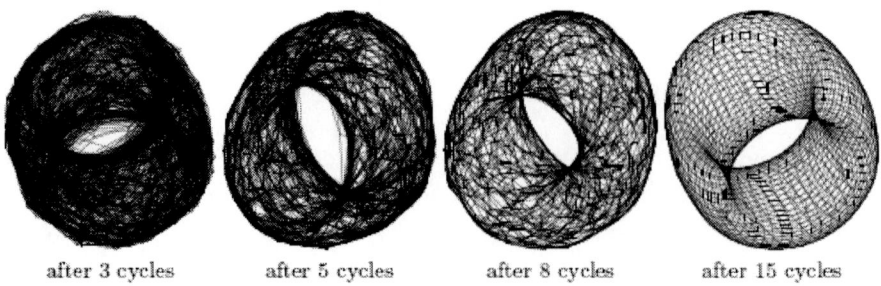

after 3 cycles    after 5 cycles    after 8 cycles    after 15 cycles

Figure 7. Constructing a torus over 50×50 nodes starting from a uniform random distribution of nodes with c = 20.

Given an arbitrary overlay network, constructing a target topology is realized via connecting all nodes to the right neighbors. T-Man's basic idea is there is a general relationship of nodes behind a given topology, which is expressed by a ranking function. The relationship between nodes could be geographical location, semantic description of stored data, storage capacity, etc.

T-Man is based on the gossip communication scheme. After initialization, each node executes the same protocol concurrently. No synchronization or coordination is needed. Nodes' running is not synchronous. The protocol consists of two threads: an active thread initiating communication with other nodes; a passive thread waiting for and processing incoming messages.

**Initialization**

        view ← rnd.view ∪ {(myAddress, myProfile)}

**Active Thread**

        **do** at a random time once in each consecutive interval of T time units
          $p$ ← selectPeer()
          *myDescriptor* ← (*myAddress, myProfile*)
          *buffer* ← merge(*view*, {*myDescriptor*})
          *buffer* ← merge(buffer, rnd.view)
          send *buffer* to $p$
          receive $buffer_p$ from $p$
          *buffer* ← merge($buffer_p$, view)

    *view* ← selectView(*buffer*)

**Passive Thread**

  **do** forever
    receive *buffer$_q$* from *q*
    *myDescriptor* ← (*myAddress, myProfile*)
    *buffer* ← merge(*view*, {*myDescriptor*})
    *buffer* ← merge(*buffer, rnd.view*)
    send *buffer* to *q*
    *buffer* ← merge(*buffer$_q$, view*)
    *view* ← selectView(*buffer*)

  As described above, each node maintains a view. The view is a set of node descriptors. Function merge(*view$_1$,view$_2$*) returns the union of *view$_1$* and *view$_2$*. In above protocol, two key functions are selectPeer() and selectView(*buffer*). Function selectPeer() uses the current view to return an address. First, it applies the ranking function to order the elements in the view. Then it returns the first descriptor that belongs to a live node. Function selectView(*buffer*) applies the ranking function to order elements in the buffer. Then it returns first *c* elements of the buffer. By using views of their current neighbors, all nodes improve their views, so that their new neighbors will be closer to the target topology. Neighbors will become closer and closer.

## 3.2. T-Chord — An Application of T-Man

### 3.2.1. Advantages of T-Chord

T-Chord efficiently bootstraps Chord from a random unstructured overlay using T-Man. It is one of most successful Chord ring building approaches in terms of thoroughness, speed, and efficiency. Simulation proved that T-Chord is able to create a perfect Chord ring in O(log(*N*)) steps where *N* is network size. It also shows optimized message latency. The generated network is immediate operable and could be handed over to the Chord protocol right away.

  T-Chord completely breaks away from the old pattern of bootstrapping structured P2P system — that is, using a jumpstart node and node joining procedure. The joining based method is very inefficient. It is unable to make nodes' bootstrapping concurrent. Nodes have to be booted one by one in a linear manner, which is very unrealistic for large network. [DBKKMSB2001] Some require booting nodes in a fixed order, which will not only need linear run time but also need complicated synchronization and coordination. Without the constraint of single jumpstart node, in T-Chord every node starts its own topology building and optimization concurrently. Furthermore, unlike many other attempts to bootstrapping Chord, T-Chord does not need any a prior configured initial network or jumpstart node.

### 3.2.2. T-Chord Protocol

T-Chord starts from a connected unstructured overlay network with a random topology. In T-Chord simulation, the unstructured random network is generated by a lightweight membership protocol called NEWCAST. [JGKS2004] Bootstrapping of T-Chord does not include node ID automatic generation. Nodes are a priori configured and unique IDs are assigned to nodes from a circular ID space. T-Man ranking function just needs minor revision to be adapted for T-Chord. In T-Man's running procedure, not only direct successor and predecessor are located as outcome of ring topology, many encountered nodes are also remembered. These buffered nodes are very useful in building Chord finger table.

### 3.2.3. Deficiencies of T-Chord

The most notable problem with T-Chord is its requirement for a priori configuration of Chord IDs. It ruins its good reputation and great prospective due to its ability to unconditionally bootstrap from arbitrary initial topology. Another short coming is its distance function, which inherited from T-Man. Its definition

$$d(u,v) = \min\{(v-u) \bmod 2^m, (u-v) \bmod 2^m\}$$

is not compatible with the distance defined in Chord, which is

$$d(u,v) = (v-u) \bmod 2^m$$

## 3.3. Ring Network

### 3.3.1. Features of Ring Network

The Ring Network (RN) protocol is an asynchronous message-passing distributed protocol, which fits well the autonomous behavior of peers in a P2P system. [SR2005] Peers do not need to be informed of any global network state. They are not required a grace leave, i.e. to assist in repairing the network topology caused by their leave.

RN protocol is not gossip based. RN uses message passing, a traditional distributed computing technique. Another notable difference is initial condition. RN requires the presence of a weakly connected initial network called minimum bootstrapping system to be able to return a Chord ring, while T-Chord can start at any condition and find any connected component. Two nodes are weakly connected means that there is a directed path between them no matter which direction the path is. For author's RAN protocol and T-Chord, differentiating weakly connected components from strongly connected components does not make much sense, since we do not have any preliminary requirement about connectivity. In addition, since most devices in MANETs support duplex mode, there is no much pragmatic significance to find this difference. RN does not specify the scale of the bootstrapping system and how the system is configured. From the Proposition 2.1 in [SR2005], we guess the bootstrapping system is a subset of all nodes to which every node is connected with at most one hop distance.

### 3.3.2. RN Protocol

The RN protocol is fully distributed. It can quickly adapt to churns in the network. All peers independently and asynchronously run a same set of procedures while they exchange asynchronous messages. Periodically each peer calls the Closer Peer Search procedure to search a closer predecessor in ID space, by which a closer successor candidate is also returned. As shown later in Section 3.3.5, authors of [SR2005] confuse successor and predecessor in the RN algorithm. But the pseudocode is still consistent and correct if we ignore the textual description.

Peers that participate in this search record information in any message they received. After collecting information returned by the predecessor search, returned by bootstrapping process, or gleaned from message propagation, each peer selects a currently closest successor. This process repeats till a complete consistent ring is formed. Local information stored by each peer includes:

- $\Gamma$: the set of current neighbors of the peer.
- $W$: the set of peers returned by Closer Peer Search.
- $B$: the set of peers that the peer has learned by the Search Monitor while propagating search request messages on behalf of other peers.
- $s$: a peer selected randomly from the current successor, and peers returned by the bootstrapping system.
- 

Three steps of the protocol are described below in more detail.

**Closer Peer Search**

Each peer $x$ periodically initiate a search for the successor candidate to which it is closer than to its current successor in the ID space. Current node first finds the closer predecessor. Current node $x$ randomly chooses a peer $s$, which is either its current successor $x.\Gamma_0$ or a peer returned by the bootstrapping system, and sends $s$ a *CloserPeerSearch* message. $s$ forwards the message to one of its neighbors to which $x$ is closest. The receiver of this request propagates this request in a similar manner. This way $x$ gets closer and closer to the target. When a receiver $u$ finds that the initiator $x$ is closer to itself than any of its neighbors, the search is terminated. $u$ then sends to $x$ the address and ID of its successor $u.\Gamma_0$, which $x$ adds to its set $x.W$.

The result of the Closer Peer Search depends on the current network topology. If the network is already in a ring topology, the search will not be really launched. Note that the search does not necessarily returns the closest node of $x$ in ID space, because the ending node of the search may have a unvisited descendent node, which is more than one hop away, and $x$ is closer to it. Furthermore, since the search is actually for a closer predecessor, it does not ensure of finding the successor to which $x$ is closest. No matter $x$ is closest to $u.\Gamma_0$ or not, since $u$ is closest to $x$, $x$ will be always between $u$ and $u.\Gamma_0$. So $u.\Gamma_0$ is a promising candidate for $x$'s successor. The frequency of this search only affects the speed of the protocol, not its correctness.

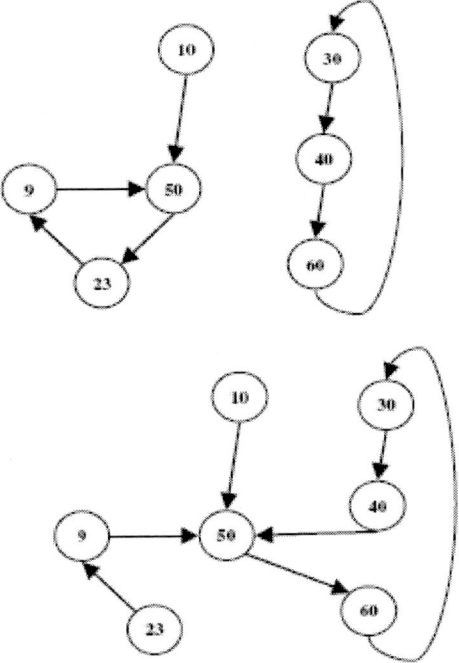

Figure 8. Closer Peer Search

Figure 8 illustrates the Closer Peer Search. Left-hand side is the starting situation; right-hand side is the ending situation. Node 50 starts this search at node 30. The search terminates at node 40, which notifies node 50 its successor 60. Node 50 then sets 60 as its new successor. Actually the exact next step for node 50 is adding node 60 to its successor candidate set $W$. To make it clearer, node 60 is assumed to be selected as new successor of node 50.

Every peer $u$ records each received Closer Peer Search message. If a search is initiated by $x \neq u$ and is terminated at $u$, then $x$ is closer to $u$ than $u.\Gamma_0$. $u$ then adds the address and ID of $x$ to its set $B$. In Figure 8, peer 40 adds 50 to its set $B$.

**Neighbor Update**

Periodically every peer $u$ checks if it has found a closer successor than its current successor $u.\Gamma_0$. It examines its current list of neighbors, a bootstrapping peer returned by the bootstrapping system, its set $W$, and its set $B$. The peer closest to $u$ from among the union of these is chosen as the new $u.\Gamma_0$. In figure 9, after $W$ and $B$ have been updated, nodes 40 and 50 update their successors as well.

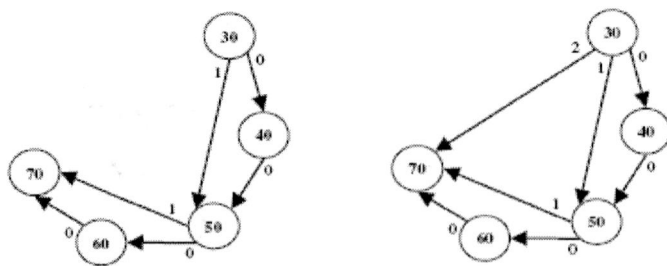

Figure 9. Neighbor update in RN

At the same time $u$ updates all its other neighbors $u.\Gamma_1$, $u.\Gamma_2$, and so on. $u$ sends a message to neighbor $u.\Gamma_i$ asking it to return the ID of $u.\Gamma_i$'s $i$th neighbor $v$. If the ID of $v$ is between $u.\Gamma_i$ and $u$, $u$ sets it as new $u.\Gamma_{i+1}$. Similar to finger table used in Chord, the purpose of such an update process is to minimize the number of hops and improve the search speed. [HGS1987] In Figure 9 peer 30 updates its third neighbor. Since the order number starts at 0, the third is actually its No.2 neighbor. It first asks its No. 1 neighbor, peer 50, for the No.2 neighbor. Peer 50 sends back 70. Peer 30 then sets peer 70 as its No.2 neighbor. This is because peer 70 is between peer 50 and peer 30 if we look at them in a ring. Eventually peer 30 has discovered a closer peer that is 4 hops away from it, using two messages.

### 3.3.3. AP Notation

AP notation is tailored pseudocode format for expressing network protocols. [Gouda1998] AP notation is very instrumental for correctness analysis. This analysis model has been proved useful and widely adopted by the distributed computing community. It ignores the execution order of interleaving of actions of nodes in a protocol by assuming arbitrarily random order. It is especially suitable for asynchronous protocols, for it expresses asynchronous protocols clearer by eliminating the need for interrupts.

In AP notation a distributed protocol consists of a series of procedures associated with nodes in a network. A node is the carrier of protocols. Data structures of a node $p$ are classified into three categories: constants, inputs, and variables, denoted by keywords **const**, **input**, and **var** respectively. The operation procedures of a protocol is put in actions section denoted by (a< $i$ >), where $i$ is the order number of procedures. Actions are delimited by two square brackets. An action is expressed in syntax

$$\text{<guard>} \rightarrow \text{<statement>}$$

The statement of an action can be executed only if the corresponding guard condition is evaluated to true. At the beginning of every round of running of a protocol, all guards of all actions of all peers are evaluated. Then only one statement of an action whose guard evaluated to true is executed. When there are more than one statements whose guards are evaluated to true, a true guarded statement is selected for execution at random. Every enabled action will eventually be executed, but the order and frequency of execution are arbitrary. RN and RAN protocols are written in AP notation.

### 3.3.4. RN Algorithm

Below is the algorithm for the RN protocol in AP notation. [SR2005]

**Peer** $u$

**const**
$T$ : set of bootstrapping peers

**input**
$w$ : a peer (successor candidate)
$x$ : peer being searched for
$c$ : index of received neighbor
$z$ : new neighbor
$s$ : a bootstrapping peer

**var**
$S$ : Set of peers
$B$ : Set of successor candidates
$W$ : Set of successor candidates
$\Gamma_i$ : $i$th neighbor

(a1) *true* $\rightarrow$
$\quad S := \{s\} \cup W \cup B \cup \Gamma$
$\quad \Gamma_0 := \text{argmin}_{k \in S}\, d(u, k)$
$\quad B := W := \varnothing$
[]

(a2) *true* →
    s:= Get random peer from $\{T \cup \Gamma_0\}$
    **send** *closerPeerSearch(u)* **to** *s*
    []

(a3) **receive** *closerPeerSearch(x)* **from** $q$ →
    **if** $x$ is closer to $u$ than any neighbor $\in \Gamma$
    **then**
        $B := B \cup \{x\}$
        **send** *successorCandidate*$(\Gamma_0)$ **to** $x$
    **else**
        **send** *closerPeerSearch(x)* **to** $\text{argmin}_{k \in \Gamma} d(k, x)$
    []

(a4) **receive** *successorCandidate(w)* **from** $q$ →
    $W := W \cup \{w\}$
    []

(a5) *true* →
    **for** each $h \in \Gamma$ **do**
        **send** *getNeighbor(index(h))* **to** $h$
    []

(a6) **receive** *getNeighbor(j)* **from** $q$ →
    **if** $\Gamma_j$ exists
    **then send** *neighbor*$(\Gamma_j, j)$ **to** $q$
    []

(a7) **receive** *neighbor(z, c)* **from** $q$ →
    **if** $\Gamma_c \leq z < u$
    **then** $\Gamma_{c+1} := z$
    **else** $\Gamma_{c+1} := \mathbf{NIL}$
    []

Note that function $\text{argmin}_{k \in S} d(u, k)$ returns a $k$, instead of $d(u, k)$ or $(u, k)$, which gives minimum $d(u, k)$.

## 3.3.5. Problems with RN

The most serious problem with RN is the minimum bootstrapping system required as a necessary condition to apply RN protocol. RN does not specify: (1) scale of the minimum bootstrapping system; (2) whether and how the minimum bootstrapping system is generated? manually or automatically by a program? from an arbitrary network topology or an a priori configured topology? (3) how many hops away from the minimum bootstrapping system is any node outside the minimum bootstrapping system? (4) how the RN is interfaced with the minimum bootstrapping system?

Second, RN is not guaranteed to converge to the ideal Chord ring within finite time. When a connected network has more nodes it is getting more difficult for RN to converge to the ideal ring. Situation in wired network is similar.

Third, as the direct reason for above problem, the basic strategy of RN in searching closer node to the target node — continuously choosing closer neighbor at each step — has no logical support at all. The common sense reasoning is against this strategy. The distribution of node IDs is totally random. The proximity of one node has nothing to do with the proximity of its children nodes. No proof of correctness of RN is presented in [SR2005].

Fourth, in [SR2005], the authors confused some basic concepts and logic. For instance, they mixed up distance from node $u$ to node $v$, i.e. $d(u, v)$ with distance from $v$ to $u$. A subsequent mix-up is the concept $u$ is closer to $v$ when $d(u, v)$ is smaller. Because the distance is directional and modulus based, suppose here the modulus is m, the following equation always holds

$$d(u,v) + d(v,u) = m$$

Obviously, by definition, when $d(u, v)$ gets smaller, distance from $u$ to $v$ becomes smaller, so $u$ is closer to $v$. At the same time, $v$ is getting farther to $u$. However, in [SR2005], the "smaller the value of $d(u, v)$ the closer $v$ is said to be to $u$." It is not just a trivial issue as chopping logic. This mistakes leads to a more serious misuse of concept in following part of the paper. For example, the loser peer search is actually a search for closer predecessor of the current node by interpreting the pseudocode of their algorithm; however, they describe it as a search for closer successor in Section 4.1, which cause a lot more confusion and logical mess-up in RN protocol and algorithm.

Next, the procedure and result of simulation of RN is not very convincing. (refer to [SR2005] Section 5) The simulator used for RN simulation is NetLogo.

[Wilensky1999] Not many models and functions for network simulation are included in libraries of NetLogo. For networking simulation the choices and possibilities are limited. In its latest version, i.e. Version 3.1, no model in the integrated library is ready for use for simulation in scenarios like RN. More important, authors of [SR2005] did not mention anything about how the simulation is implemented. No information for following questions is provided in [SR2005]: (1) whether and how the program is designed? (2) how the RN is terminated in the simulation? what is the ending condition of entire RN protocol? RN has already given the ending condition of the closer successor search, but nothing has been said about terminating the whole protocol.

Last, in simulation of RN described in [SR2005], no convergence time data or any other data about performance of RN is provided. The simulation is about the quality of Chord ring generated. Authors of [SR2005] used a concept "perfect Chord ring", however, the perfect ring does not perform best in their simulation. By definition given in Chord position paper [SMKKB2001], it is clear that there could be only one perfect Chord ring, in which all nodes in the networks are linear sorted. No other ring should be target of Chord topology construction, unless Chord is revise to a better version.

*Chapter 4*

# PREVIOUS WORKS ON STRUCTURED P2P SYSTEMS OVER MANETS

## 4.1. Special Issues on P2P Systems over MANETs

In wired networks like Internet, neighbor is defined on overlay layer and low layers such as network layer. We can say being neighbor is equivalent to knowing address. Two nodes $u$ and $v$, we say $v$ is $u$'s neighbor only if $u$ knows $v$'s network address and be able to send a message to $v$. By this definition, neighbor relation is unidirectional and not commutable. When $u$ knows $v$'s address, we have no clue if $v$ knows $u$'s address.

On the contrary, in MANETs, neighbor is only defined on lowest layer, e.g. physical layer or MAC layer. Defining layer could be expanded to network layer. In most cases, it is define by radio range. From this point of view, it has nothing to do with Chord ID space or overlay layer. In both wired networks and MANETs, the distance function is defined in the same way. From above property, a natural extension is: in MANETs, a node's neighbor set is fixed at a given time, while for a node in a wire network, it could have countless variation. Therefore, in RN protocol in Section 3.3, the neighbor update procedure can only be applied to wired networks. It is not applicable to MANETs.

For Chord or any other structured P2P systems built on wired networks like Internet, all nodes are actually connected. Even though two nodes can not connect to each other or do not know the existence of the other if they do not know the network address of the other, they are still connected. This is not the case in MANETs. Nodes in MANETs are strictly constrained by the radio range in physical layer. If there is no path from no node to another formed by neighborhood relations in a MANET, these two nodes are not reachable to each other unless their movement establish a path later. A MANET is consisted of a set of connected components, which are disjoint to each other. A component could contain only one node if the node is isolated. If a MANET has more than one component, there is no way to have one comprehensive Chord ring which includes every node like what always happen in Chord over Internet. The best scenario is we can find a Chord ring for each connected component.

Both P2P over wired networks and P2P over MANETs have proximity concerns, but in MANETs this issue is has more serious impact. The reason is still from the physical layer characteristics. A hop in MANETs is more costly than in Internet. Hence Proximity optimization has more urgent, more realistic significance in MANETs.

Substituent of IP address is necessary in MANETs for the purpose of building a P2P overlay, for example, source route in DPSR [HPD2003]. The reason is intuitive: overlay layer only makes sense or semantically correct if an underlay layer exists.

A P2P system over wired network, especially one over Internet, usually does not cover intermediate nodes of its path on the Network layer. Otherwise the P2P system may cover too many unrelated nodes. In a P2P system over a MANET, the situation is poles apart: all intermediate nodes should be included to secure connectivity on the overlay layer.

## 4.2. Cramer and Fuhrmann's Pessimistic Verdict

In Cramer and Fuhrmann's [CF2006], several serious problems could be found.

First, the whole paper is built upon some unrealistic, far-fetched assumptions. For example, they assume that all nodes can reach a common bootstrap node (which is called joint point) immediately after they power up. To make it possible, either all nodes in the MANET have to be only one hop away — which requires very small network or very powerful transmitter/receiver; or every ordinary node already has a route to that super node before power up, which is almost same as assuming that all nodes already have a pre-configured Chord successor and finger table — so the network is already initialized, why does it need bootstrapping? Another example is the assumption of single bootstrap node, which is against the definition of MANET and cause the single point of failure.

The most unrealistic assumption is at the time of power up, that is, in their own words, in the first stabilization cycle, a Chord ring has been set up and all nodes have already joined the this ring in ID space. A minor assumption, which is not serious as other assumptions, is every node knows the size of the network n. Another untenable assumption is all nodes on the ring are in a complete sequential order, from 0 to *n*.

## 4.3. Out of IP Box: Strength of RAN

In wired networks, core operations in T-Man and Ring Network are based upon IP. Traditional approach to transfer anything in wired IP networks to MANETs is replacing lower IP layers with existing routing protocols and MAC protocols of MANETs. Successful examples include DPSR [HPD2003] and Ekta [PDH2004]. If we follow this train of thoughts, the MANET version T-Man would be very complicated and inefficient, therefore not feasible in MANETs. However, the traditional approach has not been rigorously tested, even though it has been so prevalent in MANETs community. The author believes that the dominance of traditional approach is largely from the historical monopoly of IP. It is probably neither valid nor necessary.

RAN abandoned the dominant IP model and integrates the overlay layer into network and MAC layers. This avoids complicated mapping of overlay layer onto lower layers and seamlessly integrates dynamic source routing into ring-based DHT routing and tremendously reduces cost in setting up T-Man over MANETs. The resulted RAN has advantages of T-Man, T-Chord, and RN in a simplified MANET model.

The general strategy of RAN is distributed stepwise refinement. Three patterns are designed, namely distributed exhaustive pattern, virtual centralized exhaustive pattern, and random pattern. A spanning tree called component tree is used to simplify the model of connected component. A node sets itself as the root of its component tree. All nodes in the component are included in this tree. In its distributed construction procedure, the component tree goes through nodes on the fly. In two exhaustive patterns, all nodes of the tree are passed, while only nodes on one root-to-leaf path are passed in the random pattern. No node keeps its component tree in storage. Only some parts of the tree exist in memory when searching for next closer successor. This statelessness considerably increases flexibility and robustness.

Distance from node A to node B is defined as

$$(ID_B - ID_A) \bmod max$$

where *max* is a relatively huge modulo. The ring topology is determined by the successor relation among nodes of a connected component. At each step, the current successor is compared with a selected node. If the selected node has smaller distance from root than current successor, it is assigned as new successor. The process repeats till the tree is parsed.

*Chapter 5*

# RAN — AN OPTIMAL AND REALISTIC APPROACH

## 5.1. Introduction

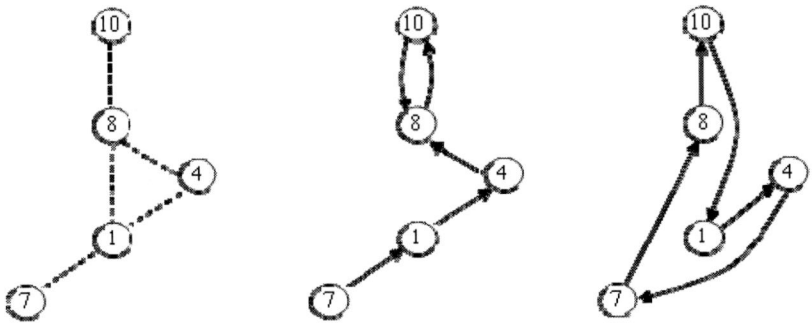

Figure 10. RAN Examples. Left is a network with only neighbor relation. Middle shows original successor relation. Right is successor relation after running RAN

Ring Ad-hoc Network (RAN) is a protocol to build a ring topology over MANETs. RAN has integrated merits of T-Man, T-Chord, and Ring Network and adapts well to MANETs. RAN is completely distributed. It uses only neighbors and local information. RAN builds an ideal ring topology for each connected component in the node ID space of a MANET. Upon this ring, ring-based P2P systems could run immediately without any stabilization. For instance, entire Chord protocol can run immediately. No stabilization is necessary unless large scale disturbance occurs. RAN integrates automatic non-IP address configuration into bootstrapping. To best of author's knowledge, it is the first successful try in the filed of bootstrapping ring-based P2P systems over MANETs.

The basic algorithm in RAN is distributed stepwise refinement. Each node treats its connected component as a tree, called component tree. All nodes in the component are included in this tree. It sets itself as the root. If the depth of a node in the tree is $i$, the node is said at level $i$. At each step, we compare the current successor with a random chosen node, all nodes in current sub tree, or all nodes in current level, depending on the pattern of the algorithm. If a chosen node in current level has shorter distance to root, we use this node as new successor. The process repeats till the tree is traversed. Here the distance function is exactly same as define by Chord, also same as that of RN.

Chord ring is determined by the successor relation among nodes in a connected component. Unlike RN, in RAN the successor of a node is not always its neighbor. If the depth of node $n$'s component tree is $p > 1$, the successor of $n$ is $n$'s neighbor only at the first round of RAN execution. As RAN runs into deeper levels, the successor may change. The distance between $n$ and its successor may be the depth of current level at most.

## 5.2. Design Goals and Assumptions

RAN is designed to achieve following goals:

- Generate an ideal Chord ring for each connected components, which will guarantee the quality of ring-based P2P systems running on the ring.
- Compatible to any MANET routing protocols, that is, routing independent.
- No any kind of a priori bootstrapping node or bootstrapping
- Pure distributed and decentralized
- Have all capability of T-Chord and RN except those incompatible with nature of MANETs
- Asynchronous, only use message passing
- Scalable to MANET size
- Good proximity and optimized for MANETs

RAN integrates automatic non-IP address configuration into bootstrapping, which is often deliberately ignored in previous approaches by assuming that an ideal IP address configuration has been a priori established from the very beginning. A non-IP node ID configuration is assumed. It generates unique random ID in structured P2P layer. No network layer address is needed. Routing in low layers uses this node ID as well.

## 5.3. Component Tree

A component tree is one spanning tree of the connected graph which is derived from a connected component. The rule of construction is:
(1) Select the searching node, which is looking up the closest successor, as the root.
(2) Add all neighbors of the root to the first level of the component tree.
(3) For all following levels, construct the next level according to the direct neighborhood relation.
(4) Delete all edges which connect a lower level node to an upper level node, no matter if the former is a descendent of latter.

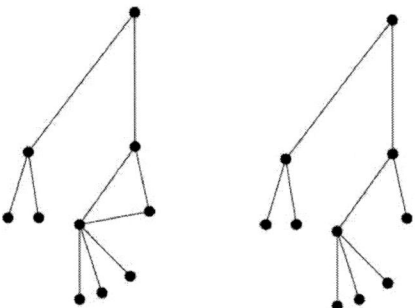

Figure 11. Convert a connected component to a component tree

For a complete component tree with $N$ nodes, uniform downward degree $k$, and the depth $d$, following equations hold.

$$N = 1 + k + k^2 + k^3 + \cdots + k^d$$

$$N = \frac{k^{d+1} - 1}{k - 1} \qquad (1)$$

Equivalently,

$$d = \log_k (kN - N + 1) - 1 \qquad (2)$$

$$k^{d+1} = kN - N + 1 \qquad (3)$$

Most performance parameters share following scenarios. With regard to number of nodes involved, we have one node parameters versus network parameters, which involve all nodes in the network. With regard to number of rounds in searching, we have one round parameters versus life-long parameters, which cover all rounds. We can mix up these two categories using simple combinations. We may have one node one round parameters, one node life-long parameters, network one round parameters, and network life-long parameters.

An ideal network is defined as a connected MANET, which has only one connected component. In an ideal network, the component tree has to be a complete tree, in which all leaf nodes are at depth $d$ or $d$-1, and all leaves at depth $d$ are toward the left. In an ideal network, one node one round message complexity $M$ is the number of messages sent in one round of searching in this node's component tree. In unicast mode, $M$ is also the number of messages received.

**Theorem 1**

In an ideal network, by symmetry, all nodes have same one node message complexity. The one round network message complexity $M^{Net}$ is the sum of one node message complexities of all nodes.

$$M^{Net} = M \times N \qquad (4)$$

**Theorem 2**

In an ideal network, the network time complexity $T^{Net}$ is equal to the one node time complexity $T$.

It applies to both one round time complexity and life-long time complexity. Obviously, all nodes run concurrently and spend same amount of time in one round of searching. Let $T_d$ be downward time, and $T_u$ upward time

$$T = T_d + T_u \qquad (5)$$

From (2) we have

$$d = O(\log N) \tag{6}$$

## 5.4. Three Patterns

To compare performance and find out intrinsic mechanism which determines the performance, we designed three patterns for RAN protocol. The primary concern is the balance between effectiveness and efficiency, to be specific, the trade-off between the completeness of generated ring and the time, message, and storage complexities of construction.

Three patterns are studied in length. Two of them are exhaustive patterns, namely, distributed exhaustive pattern and virtual centralized exhaustive pattern. In virtual centralized exhaustive pattern the searching node acts as a central controller and coordinates the searching procedure. Two exhaustive patterns use unicast in message exchange, exhaustive search at each level of component tree. Output ring is guaranteed to be ideal Chord ring for every connected component. Because it compares all node identifiers in the component, the finding the closest successor is ensured. However, this exhaustion may suffer from high cost in time, message, and storage. We need measures to mitigate the overhead. These two exhaustive patterns are equally excellent in effectiveness. Both keep 100 percent nodes of connected component in ring constructed. The distributed exhaustive pattern has better performance than the virtual centralized exhaustive pattern due to the fact that nodes in distributed exhaustive pattern only exchange messages with parents and children.

The third pattern is random pattern, which has its root in Ring Network [SR2005]. To adapt to MANETs environment, the minimum bootstrapping system is eliminated. A breadth-first search scheme is used in lieu of it. The search scheme traverses the component tree of searching node in a cascading manner to make up the poor effectiveness of RN.

### 5.4.1. Distributed Exhaustive Pattern

In the distributed exhaustive pattern, the searching node (root) sends a *getClosestCandidate* message to each of its $k$ child in its component tree. Each child concurrently forwards the message to its $k$ children at next level. At following levels, nodes keep forwarding the request message to their own children until leaf nodes are reached. Then, from leaf nodes up, the closest successor candidate of the root in the subtree is calculated at the root of the subtree and is returned to the parent node in a *closestCandidate* message. The calculation is done by comparing distances from the root to returned candidates of the children of the root of subtree. Obviously, each node sends out $k + 1$ messages except root and leaves.

$$M = (k+1)N - k^{d+1} - 1$$

Here $k^{d+1}$ is the number of skipped *getClosestCandidate* messages from $k^d$ leaf nodes, for they have no children. Similarly, 1 is the number of skipped *closestCandidate* messages from root. Plug in (3), we get

$$M = 2N - 2$$

In an ideal network with uniform downward degree $k$, the one node one round message complexity $M$ of the distributed exhaustive pattern is independent of $k$. It is only decided by the size of network $N$.

$$M^{Net} = 2N^2 - 2N$$

$$T_d = dk$$

$$T_u = d$$

$$T = d(k+1)$$

$$T = (k+1)(\log_k(kN - N + 1) - 1) = O(k \log N)$$

In an ideal network, the distributed exhaustive pattern has time complexity $O(k \log N)$.

In multicast option, each node sends out 2 messages except root and leaves.

$$M = 2N - k^d - 1$$

Plug in (3),

$$M = (1/k + 1)(N - 1)$$

$$M^{Net} = MN = (1/k+1)(N-1)N$$

When *k* is a big number,

$$M \approx N-1$$

$$M^{Net} \approx N^2 - N$$

That is,

$$M \approx O(N)$$

$$M^{Net} \approx O(N^2)$$

The messages complexity is about half of that in plain option.

The *getClosestCandidate* message needs 1 time unit to move from one level to next,

$$T_d = T_u = d$$

$$T = 2d$$

$$T = 2\log_k(kN - N + 1) - 2 = O(\log N)$$

## 5.4.2. Virtual Centralized Exhaustive Pattern

In the virtual centralized exhaustive pattern, all direct children nodes of the root form the first level. Direct children of nodes in first level form the second level, and so on. The root sends every node in current level a *getAllNeighbors* message. Then these nodes send their children set to root in an *allNeighbors* message, so root gets the next level. Then root sets the next level as new current level and repeats the same procedure till leaves are reached. This algorithm is most expensive in terms of overhead. However, it gives the root node tremendous power to control whole process upon a distributed network. Individualized services could be implemented this way.

If we count a multi-hop message as one message, both $M^{Net}$ and $M$ are independent of *k* and *d*; they only depend on *N*. The simple fact is: the root sends each node except itself a message; then each of these nodes sends a message back to the root.

$$M = 2(N-1) = 2N - 2 = O(N)$$

$$M^{Net} = 2N^2 - 2N = O(N^2)$$

It looks like same as in the distributed exhaustive pattern. However, unlike in the distributed exhaustive pattern, almost all messages have to go through multi-hops. To get precise comparison with the distributed exhaustive pattern, the per hop message complexity $M_{hop}$ should be used. At first level, there are $k$ nodes. Each needs two one hop messages. So there are $2k$ *getAllNeighbors* and *allNeighbors* messages sent to and from this level. It takes $k + 1$ time units to transfer all of them. At second level, there are $k^2$ nodes. Each needs two two-hop messages. There are $2 \times 2 \times k^2$ messages sent to and from this level. It takes $k^2 + 2$ time units to transfer all of them. Similarly, level $i$ needs $2 \times i \times k^i$ messages, which need $k^i + i$ time units to transfer.

$$M_{hop} = 2k + 4k^2 + 6k^3 + \cdots + 2dk^d = 2\sum_{i=1}^{d} ik^i$$

Suppose $S = \sum_{i=1}^{d} ik^i$

$$S = \frac{dk^{d+2} - (d+1)k^{d+1} + k}{(k-1)^2} = \frac{dk^{d+1}(k-1)}{(k-1)^2} - \frac{k^{d+1} - k}{(k-1)^2}$$

From (1),

$$N - 1 = \frac{k^{d+1} - k}{k - 1}$$

Plug it in, we have

$$S = \frac{dk^{d+1}}{k-1} - \frac{N-1}{k-1} = \frac{dk^{d+1} - N + 1}{k-1}$$

Plug in (3),

$$S = dN + \frac{d-N+1}{k-1} = O(N \log N)$$

When $N \to \infty$,

$$S \approx dN$$

$$M_{hop} = O(N \log N)$$

$$M_{hop}^{Net} = NM_{hop} = 2NS = O(N^2 \log N)$$

$$T = (k+1) + (k^2 + 2) + (k^3 + 3) + \cdots + (k^d + d)$$

$$T = N + d(d+1)/2 - 1 = O(N)$$

### 5.4.3. Random Pattern

Random pattern is the variant of RN in MANETs. The minimal bootstrapping system is replaced with the breadth-first traversal of all nodes in the component. Each round of search does begin with the starting node. The searching node gets the starting node and its route from the queue for the breadth-first traversal. It sends the node a *closerNodeSearch* message. Note that this message may need multi-hop to reach its destination when the traversal proceeds beyond neighbors of the searching node. In worst case, this message needs $d$ hops. On average, this first message needs

$$\bar{h} = \frac{k + 2k^2 + 3k^3 \cdots + dk^d}{N-1}$$

$$\bar{h} = \frac{S}{N-1} = d + \frac{d}{N-1}\frac{k}{k-1} - \frac{1}{k-1}$$

When $N \to \infty$,

$$\overline{h} \approx d$$

$$\overline{h} = O(d) = O(\log N)$$

This means when M is very big, the average hop number is close to the worst case hop number.

Along the same path of the received message, the starting node first sends back its neighbor set to the searching node to feed the traversal. Then it passes the *closerNodeSearch* message to one of its unvisited children, whose ID is closest to ID of the searching node. The receiver passes the message to one of its unvisited children, and so on, until one receiver finds that no neighbor is closer than itself. M is the number of messages exchanged between the searching node and the starting node, plus the number of messages in the search, which is $d$ in worst case and $d/2$ on average.

$$\overline{M} = \frac{d}{2} + 2 = O(\log N)$$

$$\overline{M}_{hop} = \frac{d}{2} + 2\overline{h} \approx 2.5d = O(\log N)$$

The computation time is negligible comparing to communication time, so

$$T = T_{start} + T_{search} + T_{return}$$

$T_{start}$ is the time to pass message to the starting node; $T_{search}$ is the time to find the closest node; $T_{return}$ is the time to transfer the found node ID back to the searching node.

$$T_{return} = T_{search} + T_{start}$$

On average,

$$T = 2(d + d/2) = 3d = O(\log N)$$

Due to the limit of length, detailed algorithms for three patterns are not given. Please refer to full paper if interested.

## 5.5. Three Options

Besides three patterns just described, three auxiliary options are defined to improve the efficiency, especially time complexity and message complexity. Plain option means no additional operation and the search should ends with a complete ring. Other two options are explained below.

### 5.5.1. Approximation Option

The approximation option could be applied to all three patterns. Approximation pattern does not change underlying algorithm. It works by changing the end condition of all patterns. End condition in the approximation option is much looser than normal scenario. Normally, all patterns set the ideal ring as their objective. With approximation option, a small fraction of nodes are allowed to be left out of the final rings if they are in very short line segments attached to rings.

After first running of check_rings() function in the simulator, a connected component breaks down to a ring and a set of lines which are attached to the ring at only one node. With running of the simulation, the lines gradually shrink and are absorbed by the ring. Finally with sufficient running of our simulator only ring exists.

In the plain pattern, we require that all lines are absorbed by the corresponding ring of the connected component. However, this approach becomes so resource demanding when network size increases over 100 nodes. To reduce overhead in time, message, and storage, we revise the ending condition to allow a small fraction of nodes of a component to remain in short lines. Usually the fraction is set to 10 percent, or 15 percent. This approximation tremendously reduced the complexity in time, storage, and message. The growth function of time versus network size dropped from sub exponential to linear. Similar improvement happened to the growth function of the number of sent or received messages versus network size

### 5.5.2. Multicast Option

Another option is multicast option, in which a node sends message to all direct downward neighbor nodes (its children) at next level by multicasting one message instead of unicasting multiple messages. It considerably improves time, message, and storage complexity.

However, as we mentioned above, multicast option can not be applied to any random pattern.

Please refer to Section 7 for detailed algorithm and simulation.

## 5.6. Mobility

Mobility and its complication have been one of major difficulties in MANETs. Mobility will cause leave of neighbors. It also causes repartition of connected components, which is the primary concern in topology construction. Excessive mobility may cause general failure of a MANET because it makes multi-hop communication impossible. In real world, only moderate mobility needs to de addressed.

When mobility is moderate, especially when its stable period is significantly longer than the search cycle of RAN protocols, RAN suite could solve the mobility problem with simple refreshing and dynamic variable component tree. For two exhaustive patterns, a search may miss the closest node if it just joins a node's neighbor set after this node sends out its closest successor; but second search will cover it. Refreshing is very effective and reliable. No error or exception needs to be handled. The only cost is one more round of search. For random pattern, however, the optimal successor may be missed in simple refreshing, because the traversal is not stateless. More sophisticated approach needs to be found.

Other scenarios will be addressed in the Mobile RAN protocol MRAN, which is not covered by this chapter. Please check following publications of author or contact the author by email at *weiding@ieee.org*.

*Chapter 6*

# ALGORITHMS

## 6.1. Distributed Exhaustive Pattern

### 6.1.1. Message Format

    *getClosestCandidate(originator, sender, receiver, sequence_num)*

*originator*: the ID of searching node (root of component tree)
*sender*: sender of this message, not necessarily the searching node
*receiver*: receiver of this message
*sequence_num*: a random number user to identify this message

    *closestCandidate(originator, candidate, sender, receiver)*

*originator* is the ID of searching node (root of component tree)
*candidate*: closest candidate returned
 *sender*: sender of this message, not necessarily the searching node
*receiver*: receiver of this message

    *alreadyReceived(originator, sender, receiver)*

*originator*: the ID of searching node (root of component tree)
*sender*: sender of this message
*receiver*: receiver of this message

## 6.1.2 Algorithm

**peer** $u$

**constant**
maximum: upper bound of ID

**input**
*init*: initialization flag, set to **true** at beginning
*size*: number of nodes in the connected MANET
*in-que-len*: length of incoming message queue
*out-que-len*: length of outgoing message queue

**var**
*in-queue*: incoming message queue
*out-queue*: outgoing message queue
$\Gamma$: set of one-hop neighbors
$\Gamma_0$: successor
*reply_received*: number of responses returned to *getClosestCandidate* messages sent by this node
*originator*: ID of searching node (root of component tree)
*closest_candidate(originator)*: current closest candidate for *originator*
*closest_candidate_distance(originator)*: ID space distance from *originator* to *closest_candidate(originator)*
*already_received_neighbor*: number of neighbors that already received the *getClosestCandidate* message from me (this node)
*roots_info*: map from node index of another root node to the node index of candidate of closer successor at this node for the other root node. NOT for itself.
*search_finished*: indicate the search for this node's closest successor is finished

**Library Function**
*lookfor(x, y)*: Return $x$ if $x \in y$. Return $<x, *>$ if $<x, *> \in y$. Otherwise return **NIL**.

**Action**
(a1) *init* →
    construct $\Gamma$
    **if** $\Gamma = \varnothing$

**then**
    *search_finished* := **true**
    **return**

*search_finished* := **false**
*init* := **false**
*reply_received* := 0
**for** each $h \in \Gamma$ **do**
    **send** *getClosestCandidate(u, u, h)* **to** *h*
[]

(a2) **receive** *getClosestCandidate(originator, q, u)* **from** $q \rightarrow$
    **if** *originator* $\in$ *roots_info*
    **then**
        **send** *alreadyReceived(originator, u, q)* **to** *q*
        **return**
    **else**
        *roots_info* := *roots_info* + {<*originator, q*>}
        *closest_candidate(originator)* := *u*
        *closest_candidate_distance(u)* := (*u* – *originator*) MOD *maximum*
        *already_received_neighbor(originator)* := 0
        **for** each $h \in (\Gamma - \{q\})$ **do**
            **send** *getClosestCandidate(originator, u, h)* **to** *h*
[]

(a3) **receive** *closestCandidate(originator, cd, q, u)* **from** $q \rightarrow$
    **if** *lookfor(originator, roots_info)* = **NIL**
    **then**
        *error*(*closestCandidate* message does not have a root entry)
        **exit**

    <*originator, x*> := *lookfor(originator, roots_info)*
    *reply_received* ++
    *d* := (*cd* – *originator*) MOD *maximum*
    **if** *d* < *closest_candidat_distance e(originator)*
    **then**
        *closest_candidate(originator)* := *cd*
        *closest_candidate_distance(originator)* := *d*

    **if** $u \neq originator$
    **then**
        **if** $reply\_received = |\Pi| - 1$
        **then send** $closestCandidate(originator, closest\_candidate(originator), u, x)$ **to** $x$
    **else**
        **if** $reply\_received = |\Pi|$
        **then** $search\_finished := $ **true**
[]

(a4) **receive** $alreadyReceived(originator, q, u)$ **from** $q \rightarrow$
    **if** $lookfor(originator, roots\_info) = $ **NIL**
    **then**
        $error(closestCandidate$ message does not have a root entry)
        **exit**
    $<originator, x> := lookfor(originator, roots\_info)$
    $reply\_received$ ++
    $already\_received\_neighbor$ ++
    **if** $u \neq originator$
    **then**
        **if** $reply\_received = |\Pi| - 1$
        **then send** $closestCandidate(originator, closest\_candidate(originator), u, x)$ **to** $x$
    **else**
        **if** $reply\_received = |\Pi|$
        **then** $search\_finished := $ **true**
[]

Note: For **receive** primitives, actual triggering event: message is taken out from *in-queue*.

## 6.2. Virtual Centralized Exhaustive Pattern

### 6.2.1. Message Format

The format of *getAllNeighbors* message is

$$getAllNeighbors(originator, sender, receiver, route)$$

# Algorithms

*originator*: the root node.
*sender*: sender of this message, not necessarily the searching node
*receiver*: receiver of this message
*route*: the route from the *originator*

The format of a*llNeighbors* message is

$$allNeighbors(sender, originator, neighbors, broute)$$

*originator*: *originator* of received corresponding *mGetAllNeighbors* message
*neighbors*: all neighbors except the sender of corresponding *mGetAllNeighbors* message
*broute*: route from current node to *originator*

## 6.2.2. Algorithm

**peer** $u$

**input**
*init*: initialization flag, set to **true** at beginning
*size*: number of nodes in the connected MANET
*in-que-len*: length of incoming message queue
*out-que-len*: length of outgoing message queue
*msg-rate*: message processing rate

Note: Assuming rates for incoming messages and outgoing messages are same.

**var**
*in-queue*: incoming message queue
*out-queue*: outgoing message queue
$T$: external timer, simulator by discrete counter
$L$: set of peers at the current level
$N$: set of neighbors, including all hops
*level*: current level of hops from u
$\Gamma$: set of one-hop neighbors
$\Gamma_0$: successor
$R$: $u$'s routing records for this algorithm
$r$: a route in $R$

*s*: a node in a set with smallest node ID
*AN_received*: number of received allNeighbor messages
*AN_in_queue*: number of allNeighbor messages in in_queue
*nodes_last_level*: node number in previous level
*current_completed*: if all searching is completed at current level

**Library Function**
*route*(*a*, *b*): return a route from node *a* to node *b*. Actually a route is a string or vector. *route*(*a*, *a*) returns *a*.
*distance*(*u*, *h*): RAN distance function
*reverse*(*r*): return the reverse path of route *r*

**Action**
(a1) *init* →
　　construct $\Gamma$
　　**if** $\Gamma \neq \emptyset$
　　**then**
　　　　find $\Gamma_0$
　　　　$T := 0$
　　　　*init* := **false**
　　　　$R := \emptyset$
　　　　**for** each *h* in $\Gamma$ **do**
　　　　　　$R := R + \{uh\}$
　　$N := L := \Gamma$
　　[]
Note: Here *route*(*u*, *h*) = *uh*. *uh* is a series, like string, vector in C++, ArrayList in Java.

(a2) *current_completed* →
　　*level* ++
　　*current_completed* := **false**
　　$T := 0$
　　**for** each $h \in L$ **do**
　　　　**if** route(*u*, *h*) $\in R$
　　　　**then send** *getAllNeighbors*(*u*, *u*, *h*, *route*(*u*, *h*)) **to** *h*
　　$L := \emptyset$
　　[]

(a3) **receive** *getAllNeighbors*(*o*, *q*, *u*, *r*) **from** *q* →
　　*br* := *reverse*(*r*)

        **send** a*llNeighbors*($u$, $o$, $\Gamma$ - {all nodes in $r$}, $br$) **to** $o$
        []

(a4) **receive** a*llNeighbors*($q$, $o$, $S$, $br$) **from** $q$ →
    **for** each $h$ in ($S - N$) **do**
        route($u$, $h$) := route($u$, $q$) + $h$
        $R := R +$ { route($u$, $h$)}
    $L := L + (S - N)$
    $N := N + (S - N)$
    []

(a5) **true** →
    **Local**
    *timeout_small*: lower bound of ending time
    *timeout_big*: upper bound of ending time
    *timeout_small* := 2 × *level* × *nodes_last_level* / *msg-rate*
    *timeout_big* := max{4 × *level* × *nodes_last_level* / *msg-rate*, 6 × log(*size*)}
    **if** (*AN_received* = *nodes_last_level*) **or** (($T$ >= *timeout_small*) **and** (*AN_in_queue* = 0))
                **or** ($T$ >= *timeout_big*)
    **then**
        *current_completed* := **true**
        **for** each a*llNeighbors*($q$, $u$, $S$, $br$) message still in *in-queue*
            take out a*llNeighbors*($q$, $u$, $S$, $br$)
    s := argmin$_{k \in L}$ $d(u, k)$
    **if** $d(u, s) < d(u, \Gamma_0)$
    **then** $\Gamma_0 = s$
    []

(a6) **receive** a message destined for another node →
    **send** the message to next node on the route
    []

Note: For **receive** primitives, actual triggering event is: message is taken out from *in-queue*.

## 6.3. Virtual Centralized Exhaustive Pattern with Multicast Option

### 6.3.1. Message Format

Suppose a node always sends multicast messages at low frequency, so there is no need of an out queue for sending multicast messages. Only in-queue is needed for receiving multicast messages from other nodes. We also suppose multicast messages have priority over normal messages, they could use all msg-rate to process multicast in queue if needed.

There is only one kind of multicast messages, that is, *mGetAllNeighbors*. The format of *mGetAllNeighbors* message is

*mGetAllNeighbors(originator, serial_number, sender, depth, back_route)*

*originator*: the root node.
*serial_number*: a random number used to find out later repeated coming of a same multicast message from a same originator.
*sender*: the forwarding node of the message.
*depth*: *sender*'s depth in the broadcasting tree, which is a spanning tree converted from the current connected component with root at the querying node.
*back_route*: the route back to the *originator*

The format of a*llNeighbors* message is

a*llNeighbors(sender, originator, neighbors, serial_number, depth, route)*

*originator*: *originator* of received corresponding *mGetAllNeighbors* message
*neighbors*: all neighbors except the sender of corresponding *mGetAllNeighbors* message
*serial_number*: *serial_number* of corresponding *mGetAllNeighbors* message
*depth*: depth of current node
*route*: route from current node to *originator*

### 6.3.2. Algorithm

**peer** *u*

# Algorithms

**Const**
*init*: initialization flag, set to **true** at beginning
*in-que-len*: length of incoming message queue
*out-que-len*: length of outgoing message queue

**input**
*size*: number of nodes in the connected MANET
*msg_rate*: message processing rate
*x*: the querying node at the root of the broadcasting tree
*max_depth*: maximum depth, usually log(*size*), at most *size*
*timeout*: upper bound of running time of whole procedure

**Note**: Assume rates for incoming messages and outgoing messages are same

**var**
*in_queue*: incoming message queue
*brd_in_queue*: incoming multicast message queue
*out_queue*: outgoing message queue
*T*: external timer, simulator by discrete counter
*N*: set of neighbors, including all hops
$\Gamma$: set of one-hop neighbors
$\Gamma_0$: successor
*R*: *u*'s routing records for this algorithm
*r*: a route in *R*
*s*: a node in a set with smallest node ID
*current_completed*: if all searching is completed at current level
*rnd*: a random number
*received_brdcst*: set of all received multicast messages

**Library Function**
*route*(*a*, *b*): return a route from node *a* to node *b*. Actually a route is a string or vector. *route*(*a*, *a*) returns *a*.
*distance*(*u*, *h*): RAN distance function
*reverse*(*r*): return the reverse path of route *r*

**Action**
(a1) init $\rightarrow$
    construct $\Gamma$
    if $\Gamma \neq \emptyset$

**then**
    find $\Gamma_0$
    $T := 0$
    *init* := **false**
    *rnd* := getRandomNum();
    *back_route* := route(*u*, *u*)
    **multicast** m*GetAllNeighbors*(*u*, *rnd*, *u*, 0, *back_route*)
    $R := \varnothing$
    **for** each *h* in $\Gamma$ **do**
        $R := R + \{uh\}$
    $N := \Gamma$
[]

**Note**: Here route(*u*, *h*) = *uh*. *uh* is a series, like string, vector in C++, ArrayList in Java.

(a2) **receive** m*GetAllNeighbors* →
    **if** (<*originator*, *serial_number*> $\notin$ *received_brdcst*) **and** (*depth* < *max_depth*)
        *back_route* := *back_route* + route(*sender*, *u*)
        *received_brdcst* := *received_brdcst* + {<*originator*, *serial_number*>}
        **multicast** m*GetAllNeighbors*(*originator*, *sn*, *u*, *depth* + 1, *back_route*)
        *route* := reverse(*back_route*)
        **send** a*llNeighbors*(*u*, *originator*, $\Gamma$ - {*sender*},
        *serial_number*, *depth* + 1,    *route*) **to** *originator*
[]

(a3) **receive** a*llNeighbors*(*q*, *u*, *S*, *sn*, *d*, *rt*) **from** *q* →
    **for** each *h* in (*S* − *N*) **do**
        route(*u*, *h*) := route(*u*, *q*) + *h*
        $R := R + \{$ route(*u*, *h*)$\}$
        **if** *distance*(*u*, *h*) < *distance*(*u*, $\Gamma_0$) **then** $\Gamma_0 := h$
    $N := N + (S - N)$
[]

(a4) **receive** a non-multicast message destined for another node →
    **send** the message to next node on the route
    []

(a5) T > *timeout* →

disable (a1), (a3), and (a5)
stop the whole procedure for $u$
[]

**Note**: For all **receive** primitives, actual triggering event is: message is taken out from *in-que*.

## 6.4. Random Pattern

The basic objective of random pattern is to seek high efficiency, faster convergence time, i.e. build topology faster, instead of completeness. In another word, random pattern prefers speed to the quality of ring. In this pattern, at each level of the component tree, not every node is searched like in exhaustive patterns. We pick up only one.

One way to do it is: always pick up the closest neighbor of current node. However, since node ID is assigned randomly from a huge ID space, which has nothing to do with other properties of a node, such as neighborhood. The implied strategy behind this approach — a node $x$ closer to $u$ may has neighbor even more closer to $u$ — is not tenable. However, if we use pure random selection, we will lose the ending condition. Maybe we can use the depth of searching path as an end condition.

From this point of view, RN may have chosen a better approach. Closer Peer Search in RN actually searches a node similar to the predecessor of $u$. $u$ is the current node searching closer successor. The underlying logic is: the successor of $u$'s predecessor definitely has better chance to be close to $u$.

### 6.4.1. Random Pattern Message Format

*closerNodeSearch*(*starting_node, master, serial_number, sender, depth, route, back_route*)

*starting_node*: boolean variable, indicating if this is the first *closerNodeSearch* message for the starting node.
*master*: the root node.
*sender*: the real creator and sender of the *closerNodeSearch* message.
*receiver*: the destination node.
*serial_number*: a random number used to find out later repeated coming of a same broadcast message from a same master.

*sender*: the forwarding node of the message.
*depth*: sender's depth in the broadcasting tree, which is a spanning tree converted from the current connected component with root at the querying node.
*route*: Route from sender to receiver.
*back_route*: the route back to the master.

*successorCandidate(sender, successor, receiver, serial_number, route)*

*sender*: the sender, i.e. current node.
*successor*: the successor of current node.
*receiver*: receiver of this successorCandidate message. It should be the master of its received closerPredecessorSearch message.
*serial_number*: serial_number.
*route*: the route from sender node to the master.

*newCNNeighbors(sender, receiver, serial_number, route, neighbor_set)*

*sender*: the sender of message, i.e. current node.
*receiver*: receiver of this message. It should be the master of its received *closerNodeSearch* message.
*serial_number*: serial_number.
*route*: the route back to the master.
*neighbor_set*: starting node's neighbors

## 6.4.2. Random Pattern Algorithm

### Version 5

**type** (class)
*component_node*: element of variable *component_queue*
*component_node.nodeIdx*: node index, internal expression, not node ID.
*component_node.traversed*: Indicates if a node has been traversed in current searching node's Random execution.
*component_node.in_route*: record route from current searching node to this node.

**input**
*max_depth*: maximum depth, usually log(*size*), at most *size*
*timeout*: upper bound of running time of whole procedure

**var**
*init* := **true**: mark the very beginning of algorithm
*in_queue*: incoming message queue
*out_queue*: outgoing message queue
*S*: Set of nodes
*B*: Set of successor candidates obtained while forwarding other nodes' *closerNodeSearch* messages
*W*: Set of successor candidates obtained from own *closerNodeSearch* messages
*w*: a node (successor candidate)
*x*: peer being searched for
*T*: external timer, simulator by discrete counter
$\Gamma$: set of one-hop neighbors
$\Gamma_0$: successor
*R*: *u*'s routing records for this algorithm
*s*: a node
*rnd*: a random number
*new_round*: indicates this round should stop and a new round should be started
*component_queue*: The BFS connected component queue, used as the source set to feed the *closerNodeSearch*
*dumped_component*: dumped_component contains those nodes poped out from component_queue

**Library Function**
*route*(*a*, *b*): return a route from node *a* to node *b*. Actually a route is a string or vector. *route*(*a*, *a*) returns *a*.
*distance*(*u*, *h*): RAN distance function
*reverse*(*r*): return the reverse path of route *r*
*empty*(*x*): return true if set or series x is empty, otherwise return false

**Action**

(a1) *init* →
  **Local**
  *cn*: component node
  *init* := **false**
  construct $\Gamma$
  *in_queue* := ∅
  *out_queue* := ∅

**if** $\Gamma = \emptyset$
**then**
    *new_round* := **false**
    **return**
**else**
    find $\Gamma_0$
    *new_round* := **true**
    *component_queue* := $\emptyset$
    *dumped_component* := $\emptyset$
    **for** each $h \in \Gamma$
        *cn* := *new*(*component_node*)
        *cn.nodeIdx* := *h*;
        *cn*.traversed := **true**
        *cn.in_route* := {*u*} + {*h*}
        *push_back* (*component_queue*, *cn*)
[]

(a2) *new_round* **and** (**not** *empty*(*component_queue*)) →
    *new_round* := **false**
    $B := W := \emptyset$
    *cn1* := *pop_front*(*component_queue*)
    *push_back* (*dumped_component*, *cn1*)
    *rnd* := *getRandomNum*()
    *route* := *cn1.in_route*
    *back_route* := *reverse*(*route*)
    **send** *closerNodeSearch*(**true**, *u*, *rnd*, *u*, 0, *route*, *back_route*) **to** *cn1*
[]

(a3) **receive** *closerNodeSearch*(*starting_node*, *x*, *sn*, *q*, *depth*, *route*, *back_route*))
        **from** *q* →
**Local**
*send_candidate* := **false**
*next*

**if** (*depth* >= *max_depth*)
**then**
    *send_candidate* := **true**
    *new_round* := **true**

**else if** $u$ is closer to $x$ than any $u$'s neighbor $h \in u.\Gamma$
**then**
    *send_candidate* := **true**
    $B := B \cup \{x\}$
**if** *send_candidate* = **true**
**then**
    **send** *successorCandidate*($u, \Gamma_0, x, sn, back\_route$) **to** $x$
**else**
    *next* := $\text{argmin}_{k \in \Gamma} d(k, x)$
    *rt* := *route*($u$, *next*)
    *bk_route* := *reverse*(*rt*) + *back_route*
    **send** *closerNodeSearch*(**false**, $x, sn, u, depth + 1, rt, bk\_route$) **to** *next*
**if** *starting_node* = **true**
    **send** *newCNNeighbors*($u, x, sn, back\_route, u.\Gamma$)
[]

(a4) **receive** *successorCandidate*($q, w, org, sn, route$) **from** $q \rightarrow$
    **if** $u = org$
    **then**
        $W := W \cup \{w\}$
        $S := W \cup B \cup \Gamma$
        $\Gamma_0 := \text{argmin}_{k \in S} d(u, k)$
        *new_round* := **true**
    **else send** the message **to** next node on the route
[]

(a5) **receive** *newCNNeighbors*($q, org, sn, route, nb\_set$) **from** $q \rightarrow$
    **if** $u = org$
    **then**
        **for** each $h \in nb\_set$
            **if** $h \notin (component\_queue \cup dumped\_component)$
            **then**
                *cn2* := *new*(*component_node*)
                *cn2.nodeIdx* := $h$
                *cn2.traversed* := **false**
                *cn2.in_route* := $\{u\} + \{h\}$
                *push_back*(*component_queue*, *cn2*)
[]

*Chapter 7*

# SIMULATION

To simplify programming, the simulation is based upon static network. As mentioned before, the mobile situation will be addressed in WRAN protocol, which is not covered by this chapter. It is recommended that bootstrapping is launched at a relatively less mobile setting; since mobility will change the composition of connected components. All discussion involving mobility is based upon the premise that the change in components caused by mobility disturbance should be limited to a reasonable range. The premise validates that early research could be based upon static assumption.

This simulator is not for one single connected component; instead it is for a MANET randomly generated on a $100 \times 100$ two–dimensional square. The length unit is meter. Each node is independently generated with node x coordinate and y coordinate uniformly distributed in range [0, 100], and node ID uniformly distributed in range [0, 65535]. The direct connectivity between two nodes is purely decided by their Euclidean distance and the uniform radio range for all nodes in the MANET. It is much more realistic than generating only one connected component.

Basic parameters tested include completeness, time, number of sent messages, and number of received messages. The completeness examines the effective of algorithm by checking completeness of rings generated. Time is the time used to construct rings. Two messages measure the message complexity.

## 7.1. Completeness

Completeness is the ratio of number of nodes in generated rings to number of all nodes. The simulation shows that all nodes are either in rings, or in lines. Each line is connected to one and only one ring. If more than one rings exist in one component, only nodes in the biggest ring is counted in the completeness calculation as nodes in rings. Isolated nodes are regarded as rings, so they are always counted as constructed. This is rational for some MANETs which are unfortunately initialized with considerable isolated nodes. Each connected component has one or more rings which are connected by lines. The bottom line is: at any time of the construction, even before anything is done for the construction, all nodes in same component should always remain in the same component. The construction only changes the number of rings and the nodes in rings in the component. It does not change the component, as the component is defined by the neighborhood relation among nodes, which remains identical in a static network.

**Table 1. Completeness of Algorithms**

| Network Size | 20 | 40 | 60 | 80 | 100 |
|---|---|---|---|---|---|
| Random Pattern | 0.93 | 0.705 | 0.547 | 0.395 | 0.343 |
| Distributed Exhaustive Pattern | 1 | 1 | 1 | 0.98 | 0.97 |
| Centralized Exhaustive Pattern (Plain) | 1 | 1 | 1 | 1 | 0.99 |
| Centralized Exhaustive Pattern (Approximation 0.85) | 0.94 | 0.89 | 0.87 | 0.855 | 0.86 |

## 7.2. Time

Time is defined as the algorithmic time used in one run of simulation, from beginning to end. The critical question is how the end condition is defined in simulation. As we mentioned in Section 5.5.1, normally the end condition is defined by the formation of ideal Chord ring, which is unique and fixed for a given MANET. For a pattern that needs too much time to finish, the end condition could be adapted by using the approximation option. For all combinations of patterns and options, the algorithmic time unit is set as virtually synchronized discrete time unit. In each unit, all nodes are supposed to complete the processing of $m$ incoming messages and $n$ outgoing messages. Except in multicast option, $m$

is assumed to equal to *n*. m and n are determined by the simulation parameter message processing rate.

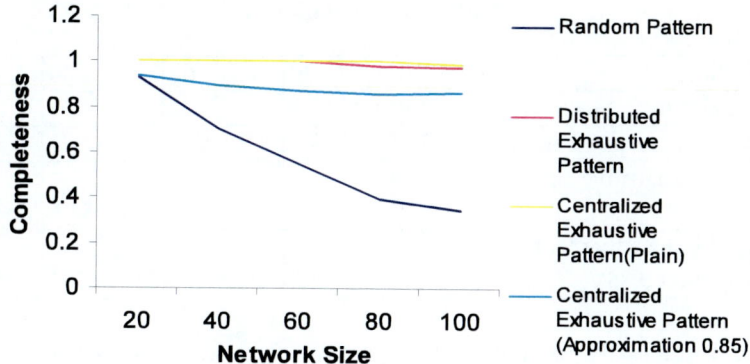

Figure 12. Completeness of algorithms.

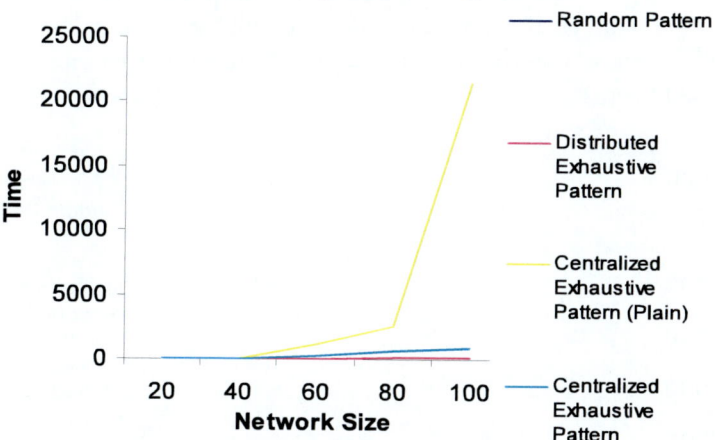

Figure 13. Time used in ring construction

**Table 2. Time used in ring construction**

| Network Size | 20 | 40 | 60 | 80 | 100 |
|---|---|---|---|---|---|
| Random Pattern | 11.8 | 15 | 32 | 63.6 | 91 |
| Distributed Exhaustive Pattern | 8.3 | 21.8 | 43 | 87.4 | 98.5 |
| Centralized Exhaustive Pattern (Plain) | 5.8 | 56.4 | 1091.4 | 2552 | 21409 |
| Centralized Exhaustive Pattern | 6 | 49.6 | 313.8 | 632 | 910 |

## 7.3. Message Complexity

**Table 3. Messages Sent**

| Network Size | 20 | 40 | 60 | 80 | 100 |
|---|---|---|---|---|---|
| Random Pattern | 72.8 | 1117.8 | 4721 | 13839 | 25117.3 |
| Distributed Exhaustive Pattern | 103.6 | 1355.1 | 5089.8 | 16077 | 26853.9 |
| Centralized Exhaustive Pattern (Plain) | 412 | 15863.4 | 162705.6 | 465754.6 | 2581659.6 |
| Centralized Exhaustive Pattern (Approximation 0.85) | 226 | 5556.8 | 62625.4 | 186703 | 364112.2 |

Messages complexity is measured by two parameters: messages sent and messages received. The message sent is defined as total number of messages sent by all nodes in the network during the simulation. The message received is defined as total number of messages received by all nodes in the network during the simulation. These messages are not user data related. They are pure control messages used to create the ring.

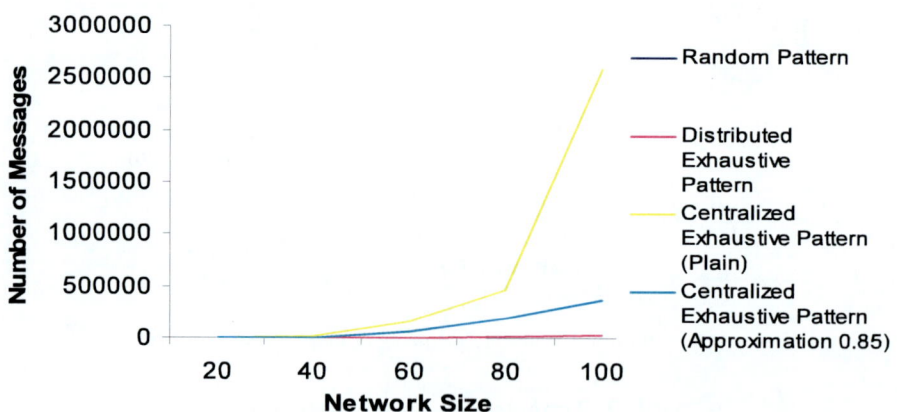

Figure 14. Messages sent

**Table 4. Messages Received**

| Network Size | 20 | 40 | 60 | 80 | 100 |
|---|---|---|---|---|---|
| Random Pattern | 72.8 | 1117.8 | 4721 | 13839 | 25117.3 |
| Distributed Exhaustive Pattern | 98 | 1312.8 | 5045 | 15938.3 | 26783 |
| Centralized Exhaustive pattern | 394.6 | 15639 | 161335.2 | 460179.4 | 2539312.2 |
| Centralized Exhaustive Pattern (Approximation 0.85) | 209.8 | 5452.2 | 61693.8 | 183749.6 | 358354 |

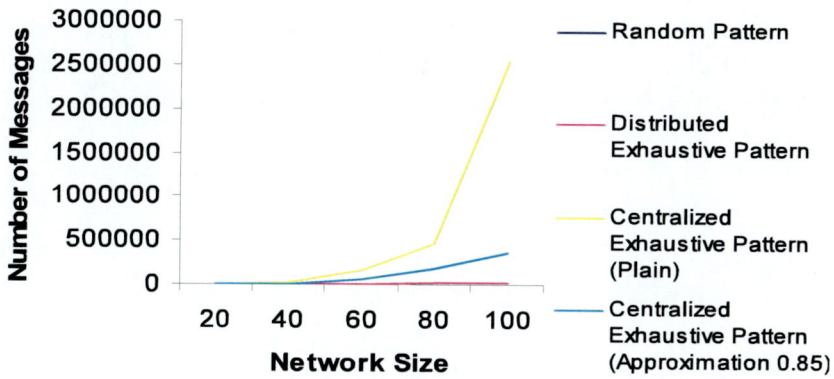

Figure 15. Messages received.

## 7.4. Analysis of Simulation Results

Obviously, the winner is the distributed exhaustive pattern. It shows perfect effectiveness, at the same time and unlike other two patterns, it has no serious side effect. Other two, however, suffer from different fatal problems. For random pattern, it is the effectiveness. For virtual centralized exhaustive pattern, it is efficiency.

As shown in preceding sections, it is clear that in RAN family, random pattern has the best overhead cost in both time complexity and message complexity. However, it is also the worst approach in terms of quality of ring constructed. The reason is behind its searching strategy. The closest first criterion does not make sense in a pure stochastic uniform distribution of ID space. The closest successor could be hidden anywhere in the component tree. It could be child of any node. It may be child of current closest successor, or child of current farthest successor, or child of any other node. The lesson is: in face of such complete randomness, exhaustion in search is necessary. It has been illustrated clearly. Exhaustive approaches almost always returns the best rings, unless we intentionally prohibit it from doing so with an approximation.

The simulation demonstrates poor efficiency of centralized approach. The virtual centralized exhaustive pattern is worst in terms of cost in time and message. On the other hand, its twin approach, the distributed exhaustive pattern shows tremendous divergent performance. It proved that exhaustion is not necessarily a synonym of expenditure. The distributed exhaustive pattern has almost same efficiency as the random pattern.

The third enlightenment is at a high level of abstraction, it is kind of philosophy. As the author always advocate, the benefit of decentralization has shown by the distributed exhaustive pattern. This result also raises a question: could centralization be implemented above decentralized infrastructure?

*Chapter 8*

# CONCLUSION

This chapter introduces a novel approach for bootstrapping P2P overlay ring over MANETs. Originally it could be used to all ring-based P2P systems, like Pastry and recent Virtual Ring Routing. This approach benefits from successful P2P topology construction methods in wired networks. To the best of the author's knowledge, this approach is the first successful attempt in this field.

RAN protocol is proposed for ring topology construction. RAN builds perfect ring in P2P ID space. It integrates upper layer into lower layer. No underlying routing protocol is needed. Ring-based P2P systems could be immediately put into normal operation upon the ring. RAN includes a variety of algorithms. Pros and cons of these algorithms are shown, both in theory and in simulation. Simulation shows that the distributed exhaustive pattern is the best in terms of effectiveness and efficiency.

This is a new area of study; many questions are unanswered; many research topics could be developed. Here we only give one example. As explanation for poor effectiveness of both Ring Network and RAN-Random, the very idea to keep tracing the closest node at every round of closer successor search, does not yield best successor, nor does using its other varieties like searching for closest predecessor. Mathematical analysis and simulation results both show the weakness of this approach: its guideline of finding closest node limits its range of choice. To improve this but not going to another extreme of exhaustive search, another approach could be tried. That is: keep the random itinerary but discard the closest standard. However, a follow-up question would be immediately raised, that is: without the closet criterion, what can be our end condition? Simplest answer is search depth or search time, or quality of returned node. In this direction, we guess the biggest gold mine may be under the way. It could be the most prospective follow-up research for RAN.

# REFERENCES

[BEGH2004] M. Brahami, Patrick Th. Eugster, Rachid Guerraoui, Sidath B. Handurukande: BGP-Based Clustering for Scalable and Reliable Gossip Broadcast. Global Computing 2004, pp. 273-290, Rovereto, Italy, 2004

[BRAN2006] TC (Technical Commitee) BRAN, "ETSI HIPERLAN/2 Standard," http://portal.etsi.org/radio/HiperLAN/HiperLAN.asp, June 2006 (Last updated: 2006-06-05 15:13:24).

[Boleng2002] J. Boleng, "Efficient Network Layer Addressing for Mobile Ad Hoc Networks," Proceedings of the International Conference on Wireless Networks (ICWN'02), pP 271–277, Las Vegas, NV, June 2002.

[Borg2003] Joseph Borg, *"A Comparative Study of Ad Hoc & Peer to Peer Networks"*, M.S. Thesis, University College London, August 2003.

[CAG2005] S. Cheshire, B. Aboba, and E. Guttman, *"Dynamic configuration of IPv4 link-local addresses,"* Proposed Standard, Internet Engineering Task Force, draft-ietf-zeroconf-ipv4-linklocal, May 2005.

[CCL2003] Imrich Chlamtac, Marco Conti, Jennifer J.-N. Liu, *"Mobile ad hoc networking: imperatives and challenges"*, Ad Hoc Networks 1, pp 13–64, Elsevier Press, 2003.

[CCNOR2006] Matthew Caesar, Miguel Castro, Edmund B. Nightingale, Greg O'Shea, and Antony Rowstron, *"Virtual Ring Routing: Network Routing Inspired by DHTs,"* Proc. ACM SIGCOMM 2006.

[CF2006] Curt Cramer and Thomas Fuhrmann, *"Bootstrapping Chord in Ad Hoc Networks: Not Going Anywhere for a While,"* Fourth Annual IEEE International Conference on Pervasive Computing and Communications Workshops (PERCOMW'06), pp. 168-172, Pisa, Italy, 2006.

[CWLG1997] C.-C. Chiang, H.K. Wu, W. Liu, M. Gerla, *"Routing in clustered multihop, mobile wireless networks with fading channel,"* Proceedings of IEEE SICON97, pp. 197–211, April 1997.

[Clip2] Clip2, *"The Gnutella Protocol Specification v0.4,"* Document Revision 1.2, http://www9.limewire.com/developer/ gnutella_protocol_0.4.pdf.

[Cohen2003B] 33. Bram Cohen, *"BitTorrent Economics Paper,"* May 2003 http://bitconjurer.org/BitTorrent/bittorrentecon.pdf

[Cohen2003I] Bram Cohen, *"Incentives Build Robustness in BitTorrent,"* May 2003. http://www.bittorrent.com/bittorrentecon.pdf

[DBKKMSB2001] Frank Dabek, Emma Brunskill, M. Frans Kaashoek, David Karger, Robert Morris, Ion Stoica, Hari Balakrishnan, *"Building Peer-to-Peer Systems with Chord, a Distributed Lookup Service,"* In the Proceedings of the 8th Workshop on Hot Topics in Operating Systems (HotOS-VIII), Schloss Elmau, Germany, May 2001.

[DVH2003] Nitin Desai, Varun Verma and Sumi Helal, *"Infrastructure for Peer-to-Peer Applications in Ad-Hoc Networks"*, 2nd International Workshop on Peer-to-Peer Systems(IPTPS), Berkeley, CA, February 2003.

[EE2000] J. Elson and D. Estrin. An Address-free Architecture for Dynamic Sensor Networks. Technical Report 00-724, Computer Science Department, USC, January 2000.

[FL2001] James A. Freebersyser, Barry Leiner, *"A DoD perspective on mobile ad hoc networks"*, in: Charles E. Perkins (Ed.), Ad Hoc Networking, pp. 29–51, Addison Wesley, Reading, MA, 2001.

[Gast2002] M. S. Gast, *"802.11 Wireless Networks – The Definitive Guide,"* O'Reilly & Associates, California 2002.

[Gouda1998] M. G. Gouda, *"Elements of Network Protocol Design,"* John Wiley and Sons, 1998.

[HDN2003] R. Hinden, S. Deering, E. Nordmark, *"RFC3587: IPv6 Global Unicast Address Format,"* Proposed Standard, Internet Engineering Task Force, August 2003.

[HDVL2003] Sumi Helal, Nitin Desai, Varum Verma and Choonhwa Lee, *"Konark - A Service Discovery and Delivery Protocol for Ad-Hoc Networks"*, Proceedings of the Third IEEE Conference on Wireless Communication Networks(WCNC), New Orleans, Louisiana, March 2003.

[HGRW2006] T. Heer, S. Gotz, S. Rieche, and K. Wehrle, *"Adapting Distributed Hash Tables for Mobile Ad Hoc Networks,"* Proceeding of Fourth Annual IEEE International Conference on Pervasive Computing and Communications Workshops, pp.173 – 178, 2006.

[HGS1987] W. D. Hillis, J. Guy, and L. Steele, "*Data Parallel Algorithms*," *Communication ACM*, **30**(1), pp.78–78, 1987.

[HPD2003] Y. C. Hu, H. Pucha, and S. M. Das, "*Exploiting the Synergy between Peer-to-Peer and Mobile Ad Hoc Networks*," Proceedings of HotOS-IX: Ninth Workshop on Hot Topics in Operating Systems, Lihue, Kauai, Hawaii, May 2003.

[Henson2003] Val Henson, "*An Analysis of Compare-by-Hash*," Proceedings of the 9th Workshop on Hot Topics in Operating Systems, Lihue, Hawaii, May 2003

[Heritage2000] American Heritage. "*The American Heritage Dictionary of the English Language*", Fourth Edition, Houghton Mifflin Company, Boston, MA, January 2000.

[Ivkovic2001] Igor Ivkovic, "*Improving Gnutella Protocol: Protocol Analysis and Research Proposals*," Prize-Winning Paper for LimeWire Gnutella Research Contest, September 2001

[JB2005] M. Jelasity and O. Babaoglu, "*T-Man: Gossip-based overlay topology management*," In Engineering Self-Organising Applications (ESOA'05), 2005.

[JGKS2004] M. Jelasity, R. Guerraoui, A.-M. Kermarrec, and M. van Steen, "*The Peer Sampling Service: Experimental Evaluation of Unstructured Gossip-Based Implementations*," In Middleware 2004, volume 3231 of Lecture Notes in Computer Science, pp. 79–98, Springer-Verlag, 2004.

[JHB2001] K. Jenkins, K. Hopkinson, and K. Birman, "*A Gossip Protocol for Subgroup Multicast*," In International Workshop on Applied Reliable Group Communication (WARGC), April 2001.

[JM1996] D.B. Johnson, D.A. Maltz, "*Dynamic source routing in adhoc wireless networks*," in: T. Imielinski, H. Korth (Eds.), Mobile Computing, pp. 153–181, Kluwer Academic Publishers, Dordrecht, 1996.

[KLW2003] A. Klernm, C. Lindemann and O. Waldhorst, "*A special Purpose Peer-to-Peer File Sharing System for Mobile Ad Hoc Networks*", Proc. Workshop on Mobile Ad Hoc Networking and Computing (MADNET 2003), Sophia-Antipolis, France, pp 41-49, March 2003.

[Kaashoek2003] Frans Kaashoek, "*Peer-to-peer computing research: a fad?*" 2003 http://project-iris.net/talks/dht-toronto-03.ppt

[Kortuem2001] Gerd Kortuem. "*Proem: A Peer-to-Peer Computing Platform for Mobile Ad-hoc Networks*", Online proceedings of Advanced Topic Workshop in Middleware for Mobile Computing, November 2001.

[LJLQC2004] Sei-yon Lee, Ju-wook Jang, Kyung-Geun Lee, Lan Quan, Tae-kyoung Cho, "*A Peer-to-Peer Search Scheme over Mobile Ad hoc*

*Networks*," (ISPC) International Scientific-Practical Conference 2004, Institute of Mathematics of National Academy of Sciences (IM NAS, Bishkek, Kyrgyz Republic), 2004.

[LLS2004] Mei Li, Wang-Chien Lee, Anand Sivasubramaniam, "*Efficient peer to peer information sharing over mobile ad hoc networks*," the Second WWW Workshop on Emerging Applications for Wireless and Mobile Access (MobEA'04), New York City, NY, May 2004.

[LMM2000] Meng-Jang Lin, Keith Marzullo, and Stefano Masini, "*Gossip versus Deterministically Constrained Flooding on Small Networks*," Proceedings of the International Symposium on Distributed Computing (DISC), Toledo, Spain, October 2000.

[LW2002] C. Lindemann and O. Waldhorst, "*A Distributed Search Service for Peer-to-Peer File Sharing in Mobile Applications*," Proceeding of. 2nd IEEE Conference on Peer-to-Peer Computing, 2002.

[MDMD2001] Archan Misra, Subir Das, Anthony McAuley, and Sajal K. Das, "*Autoconfiguration, Registration, and Mobility Management for Pervasive Computing*", IEEE Personal Communication, pp 24-31, August 2001

[MG1996] S. Murthy and J.J. Garcia-Luna-Aceves, "An Efficient Routing Protocol for Wireless Networks", ACM Mobile Networks and Application, Special Issue on Routing in Mobile Communication Networks, pp. 183-97, Oct. 1996.

[MMA2000] K-C Mei, R Mathur, S. K. Agarwal, "*Gossip Style Data Stability in Networks*," Project Report, Boston University, December 2000.

[MP2002] Mansoor Mohsin and Ravi Prakash, "*IP Address Assignment in Mobile Ad Hoc Networks,*" Proceedings of IEEE MILCOM, September 2002.

[Muthusamy2003] Vinod Muthusamy, "*An Introduction to Peer-to-Peer Networks*," October 2003. http://www.eecg.toronto.edu/~jacobsen/mie456/slides/p2p-mie.pdf

[Naugle1998] Matthew Naugle, "*Illustrated TCP/IP – A Graphic Guide to the Protocol Suite*," John Wiley & Sons, Inc., November 1998.

[OSL2003] B. Oliveira, I.G. Siqueira, A.A. Loureiro, "*Evaluation of ad-hoc routing protocols under a peer-to-peer application*," IEEE Wireless Communication and Networking Conference, 2003

[OSMLWN2005] Leonardo B. Oliveira, Isabela G. Siqueira, Daniel F. Macedo, Antonio A. F. Loureiro, Hao Chi Wong, Jose M. Nogueira, "*Evaluation of Peer-to-Peer Network Content Discovery Techniques over Mobile Ad Hoc Networks*," IEEE International Symposium on a World of Wireless, Mobile and Multimedia Networks (WOWMOM'05), pp. 51-56, Taormina, Italy, June 2005.

[PB1994] C.E. Perkins, P. Bhagwat, "*Highly dynamic destination-sequenced distance-vector routing (DSDV) for mobile computers,*" Computer Communications Review (October 1994), pp. 234–244, 1994.

[PC1997] V.D. Park, M.S. Corson, "*A highly adaptive distributed routing algorithm for mobile wireless networks,*" in: Proceedings of INFOCOM 97, April 1997.

[PDH2004] Himabindu Pucha, Saumitra M. Das, Y. Charlie Hu, "*Ekta: An Efficient DHT Substrate for Distributed Applications in Mobile Ad Hoc Networks,*" Sixth IEEE Workshop on Mobile Computing Systems and Applications, pp. 163-173, 2004.

[PMWBS2001] C. Perkins, J. Malinen, R. Wakikawa, E. Belding-Royer, and Y. Sun, "*IP Address Autoconfiguration for Ad Hoc Networks*", IETF Internet-Draft, November 2001 (work in progress).

[PR1999] C.E. Perkins, E.M. Royer, "*Ad-hoc on-demand distance vector routing,*" in: Proceedings of 2nd IEEE Workshop on Mobile Computing Systems and Applications, February 1999.

[PS2001] M. Papadopouli and H. Schulzrinne, "*Effects of power conservation, wireless coverage and cooperation on data dissemination among mobile devices,*" Proceedings of ACM Interational Symposiumon Mobile Ad Hoc Networking and Computing (MobiHoc), pp 117–127, October 2001.

[RD2001] A. Rowstron and P. Druschel, "*Pastry: Scalable, distributed object location and routing for large-scale peer-to-peer systems,*" Proceeding of IFIP/ACM International Conference on Distributed Systems Platforms (Middleware), pp 329-350, Heidelberg, Germany, November 2001.

[RFHKS2001] S. Ratnasamy, P. Francis, M. Handley, R. Karp, S. Shenker, "*A Scalable Content-Addressable Network,*" Proceedings of the SIGCOMM, pp 161-172, 2001

[RR2002] R. Ramanathan and J. Redi, "*A Brief Overview of Ad Hoc Networks: Challenges and Directions,*" IEEE Communications, 50th Anniversary Commemorative Issue, pp. 20-22, May 2002.

[RS1998] Ram Ramanathan, Martha Steenstrup, "*Hierarchically organized, multihop mobile wireless networks for quality of service support*", Mobile Networks and Applications, 1998 No.3, pp101–119, 1998.

[RT1999] Elizabeth M. Royer, Chai-Keong Toh, "*A Review of Current Routing Protocols for Ad Hoc Mobile Wireless Networks*", IEEE Personal Communications, Vol. 6, No. 2, pp. 46-55, April 1999.

[SB2003] Y. Sun and E. M. Belding-Royer, "*Dynamic Address Configuration in Mobile Ad hoc Networks,*" Technical Report 2003-11, Computer Science Department, UCSB, March 2003.

[SMKKB2001] Ion Stoica, Robert Morris, David Karger, M. Frans Kaashoek, and Hari Balakrishnan, "*Chord: A Scalable Peer-to-peer Lookup Service for Internet Applications*," In Proceeding of ACM SIGCOMM 2001, pp. 149-160, San Diego, CA, August 2001.

[SR2005] A. Shaker and D. S. Reeves, "*Self-stabilizing structured ring topology P2P systems*," Technical Report 2005-25, Department of Computer Science, N.C. State University, 2005.

[VM2003] John Viega and Matt Messier, "*Secure Programming Cookbook for C and C++: Recipes for Cryptography, Authentication, Input Validation & More*," 1 edition, O'Reilly Media, Inc, July 2003

[Vaidya2002] N. H. Vaidya, "*Weak Duplicate Address Detection in Mobile Ad Hoc Networks*", Proceedings of the ACM International Symposium on Mobile Ad Hoc Networking and Computing (MobiHoc' 02), pp 206–216, Lausanne, Switzerland, June 2002.

[Wilensky1999] U. Wilensky, "*NetLogo*," http://ccl.northwestern.edu/netlogo/, Center for Connected Learning and Computer-Based Modeling, Northwestern University, Evanston, IL., 1999.

[ZNM2003] H. Zhou, L. Ni, and M. Mutka, "*Prophet Address Allocation for Large Scale MANETs*," Proceedings of the IEEE Conference on Computer Communications (INFOCOM 2003), San Francisco, CA, March 2003.

# INDEX

## A

accuracy, 12
achievement, 2
ad hoc network, 71, 72, 74
ad hoc networking, 71
advertising, 2
algorithm, 11, 18, 23, 27, 29, 36, 41, 45, 46, 51, 55, 59, 63, 75
assumptions, 2, 3, 32, 33
attacks, 8
availability, 8

## B

bandwidth, 5, 6, 8
behavior, 23
Bluetooth, 1
buffer, 20, 21

## C

C++, 52, 56, 76
candidates, 27, 40, 59
carrier, 26
children, 29, 39, 40, 41, 44, 45
chopping, 29
clients, 6
clustering, 2

communication, 5, 9, 17, 18, 20, 44, 46
communication overhead, 18
community, vii, 17, 26, 33
complexity, 38, 40, 41, 42, 45, 63, 66, 67
components, 23, 32, 36, 46, 63
composition, 63
computation, 11, 44
computing, 5, 6, 73
concurrency, 3
configuration, 2, 22, 35, 36, 71
confusion, 29
connectivity, 23, 32, 63
conservation, 75
construction, vii, 2, 3, 4, 30, 33, 37, 39, 46, 64, 65, 69
consumers, 5, 6
control, vii, 41, 66
convergence, 17, 18, 30, 57
costs, 2
CPU, 5, 6
cycles, 5, 6

## D

data structure, 6
database, 6, 8
decentralization, vii, 1, 68
defects, 6
definition, 22, 29, 30, 31, 32
delivery, 12

# Index

democracy, 6
denial, 8
distributed applications, 17
distributed computing, 7, 17, 23, 26
distribution, 6, 10, 11, 12, 20, 29, 67
dominance, 5, 33
draft, 71
dynamic systems, 18

## E

email, 46
English Language, 73
enthusiasm, vii
environment, 39
equality, 5
evolution, 18
execution, 26, 27, 36, 58

## F

failure, 5, 6, 8, 32, 46
family, 5, 67
fault tolerance, 5, 7
flexibility, 18, 33
flooding, 8, 18
France, 73
freedom, 5

## G

generation, 22
Germany, 72, 75
goals, 36
gold, 70
gossip, 17, 18, 20, 23
government, 10
graph, 37
grid computing, 7
growth, 45

## H

Hawaii, 73
homogeneity, 1
Hops, 12
host, 5, 9
hybrid, 8

## I

implementation, 9
information sharing, 74
infrastructure, 1, 68
interface, 9, 10
interval, 20
IP address, 5, 7, 13, 32, 35, 36
IP networks, 2, 5, 33
iris, 73
Israel, 10
Italy, 71, 74

## J

Java, 52, 56

## L

latency, 21
links, 18
load balance, 12
Louisiana, 72

## M

MAC protocols, 33
MANETs, v, vii, 1, 2, 3, 4, 23, 31, 32, 33, 35, 36, 39, 43, 46, 64, 69, 76
mapping, 33
market, 1
measures, 19, 39
membership, 22
memory, 6, 33

message passing, 23, 36
messages, 11, 18, 20, 23, 24, 26, 38, 39, 40, 41, 42, 44, 45, 48, 51, 52, 54, 55, 59, 63, 64, 66
Microsoft, 9
mobile device, 75
mobility, 46, 63
models, 30
modulus, 29
monopoly, 33
movement, vii, 32
music, 5, 7

## N

naming, 9
network, vii, 1, 2, 3, 6, 7, 9, 10, 11, 12, 17, 18, 20, 21, 22, 23, 24, 26, 29, 30, 31, 32, 33, 35, 37, 38, 40, 41, 45, 63, 64, 66
networking, vii, 2, 30
nodes, 2, 6, 7, 8, 9, 10, 11, 12, 18, 19, 20, 21, 22, 23, 25, 26, 29, 30, 31, 32, 33, 34, 36, 37, 38, 39, 40, 41, 42, 43, 45, 48, 51, 52, 53, 54, 55, 59, 63, 64, 66

## O

one dimension, 19
optimization, 22, 32
overlay, vii, 2, 3, 5, 9, 20, 21, 22, 31, 32, 33, 69, 73

## P

parameter, 8, 65
parents, 39
partition, 3
passive, 20
peers, 5, 7, 8, 23, 24, 27, 51
poor, 6, 12, 39, 68, 69
power, 1, 6, 32, 41, 75
preference, 18
privacy, 5
probability, 1, 12

producers, 5
program, 17, 29, 30
programming, 63
propagation, 23
protocol, vii, 3, 4, 17, 18, 20, 21, 22, 23, 24, 26, 27, 29, 30, 31, 35, 39, 46, 63, 69, 72

## Q

quality of service, 75
query, 8, 9

## R

radio, 31, 63, 71
range, 17, 19, 31, 63, 69
reasoning, 29
recovery, 3, 17
redundancy, 18
refining, 17
relationship, 5, 20
relationships, 2, 7
reputation, 22
resources, 5, 6
returns, 9, 21, 24, 28, 52, 55, 59, 68
rings, vii, 45, 63, 64, 68
robustness, 2, 17, 33
routing, vii, 2, 3, 7, 10, 11, 33, 36, 51, 55, 59, 69, 73, 74, 75

## S

scaling, 6
search, 7, 8, 9, 11, 14, 23, 24, 25, 26, 29, 30, 39, 43, 44, 45, 46, 48, 49, 50, 68, 69
searches, 7, 57
searching, 29, 33, 37, 38, 39, 43, 44, 47, 48, 51, 52, 55, 57, 58, 67, 69
second generation, 9
seed, 2
self-organization, 17
separation, 2
sharing, 2, 5, 6, 7, 8, 9, 10

simulation, 18, 22, 29, 30, 45, 46, 63, 64, 66, 68, 69
slaves, 6
Spain, 74
speed, 18, 21, 24, 26, 57
stabilization, vii, 3, 32, 35
storage, 5, 6, 8, 9, 20, 33, 39, 45
subgroups, 7
substitution, 2
substrates, 10
Sun, 75
Switzerland, 10, 76
symmetry, 38
synchronization, 20, 22

## T

targets, 17
terminals, 6
threat, 6
time use, 63, 64
topology, vii, 2, 3, 7, 10, 17, 18, 20, 21, 22, 23, 24, 29, 30, 34, 35, 46, 57, 69, 73, 76
topology management, 73
torus, 10, 20
tracking, 11
trade, 39
trade-off, 39
traffic, vii

transformation, 19
transplantation, vii, 1
transportation, 18

## U

uniform, 11, 20, 37, 40, 63, 67
user data, 66

## V

variables, 26
variation, 31
vector, 19, 52, 55, 56, 59, 75
versatility, 18

## W

weakness, 69
wireless networks, 72, 73, 75
World Wide Web, 6
WWW, 74

## Y

yield, 69